ROOFING & SIDING

BY THE EDITORS OF SUNSET BOOKS

AND SUNSET MAGAZINE

LANE PUBLISHING CO.

MENLO PARK, CALIFORNIA

Supervising Editor: **Barbara G. Gibson**

Research and Text: **Lee Foster**
          Donald W. Vandervort
          Elton Welke

Design: **Joe di Chiarro**

Illustrations: **Mark Pechenik**

Cover: Photograph by Ells Marugg.
     Cover design by Zan Fox.

Editor, Sunset Books: Elizabeth L. Hogan

Fifth printing March 1989

We gratefully acknowledge . . .

. . . the many individuals who so
generously contributed their time,
talents, and expertise to the
preparation of this book.

We extend special thanks to
Hilary Hannon for her help in
uncovering ideas you see presented
in the color section of the book.

For their consulting help, we
also would like to thank Philip
Benfield of the American Plywood
Association; Roy Burgk of Center
Wholesale Company; Albert J.
Coyle, Koppers Company, Inc.;
J. G. Graef, Jr., Johns-Manville Sales
Corporation; W. A. Growdon, Gory
Associated Industries, Inc.;
W. W. Harrington, CertainTeed
Corporation; James M. Hay, The
Flintkote Company; Gary L. Hansel,
Masonite Corporation; Dick Isaacs,
Cleasby Manufacturing Company;
Keith Kersell, California Redwood
Association; Henry Koopmann,
Bird & Son; Dick Lane, Rocky
Mountain Supply, Inc.;
S. M. Lewis, Asphalt Roofing
Manufacturers Association; Patrick
L. Malley; Tom Paarmann,
California Tile, Inc.; John H. Place,
Euro-Cal Trading, Inc.; Walter F.
Pruter, National Tile & Panel
Roofing Manufacturers Institute,
Inc.; Marshall Ritchie and Franklin
C. Welch, Red Cedar Shingle &
Handsplit Shake Bureau; Selwyn
Shmitt, Lumaside, Inc.; Bud
Spangler, Alcoa Building Products,
Inc.; and Conni Vogelsang, Ford
Wholesale Company, Inc.

Photographers

Edward Bigelow: 20 top. **Lee Foster:** 11
top right. **Barbara Gibson:** 26 bottom
left. **Johns-Manville:** 6 top left and right,
8 bottom. **Steve W. Marley:** 2, 5 top, 7,
8 top, 9 left, 13 bottom left and right, 14,
15 left and top right, 17 top left, 18 left,
19 top, 21, 23 right, 24 left, 27 top and
bottom left and bottom right. **Ells
Marugg:** 16 right. **Jack McDowell:** 3, 4,
5 bottom, 6 bottom left and top and
bottom center, 9 top and bottom right,
10, 11 left and bottom right, 12 top left,
top and bottom center and top and
bottom right, 13 right, 15 bottom right,
16 left and center, 17 right and bottom,
18 right, 19 bottom, 20 bottom, 22,
23 top and bottom left, 24 right, 25 top
right and bottom, 26 all but bottom left,
27 top right, 28, 29, 31, 32. **Reynolds
Aluminum:** 30. **Shakertown Fancy Cuts
Shingles:** 12 bottom left. **Simpson
Timber Company:** 25 top left.

Architects and designers

Clark Bloomfield: 8 top; **James E. Bryant:**
12 top right; **Tom Butt of Interactive
Resources:** 24 right; **Roy E. Colbert:** 14;
Richard Crandall: 15 left; **Charles W.
Delk:** 20 bottom; **Howard G. Ding:** 17 top
left; **Bill Empey:** 27 top left; **David S.
Gould Design/Planning:** 24 left; **Dennis R.
Holloway and University of Minnesota
Environmental Design Studio:** 20 top;
John F. Janiga: 9 bottom right; **Sanford M.
Kellogg:** 9 top right, 32; **Jerry
Langkammerer:** 26 right; **J. Lamont
Langworthy:** 21 left; **Robert and Laura
Migliori:** 29; **Nagle/Hartray & Associates:**
5 bottom, 28 right; **Newman/Lustig &
Associates:** 23 top left; **Fred Polito:** 4,
22 right; **Richard L. Price:** 17 right; **Ray
and Jacqueline Rossi of Designed
Environ, Inc.:** 18 left and 19 top; **Lisa L.
and H. Michael Schneider:** 17 bottom;
Jerry Shoffner: 27 bottom left; **Carey
Smoot:** 14; **Jeff Waymack of Waldron,
Huston & Barber Architects:** 16 right;
Zelma Wilson and Associates: 5 top, 7;
Winston & May: 25 bottom.

CONTENTS

YOUR HOUSE

IDEAS FOR ROOFING

EXTERIOR

A N D S I D I N G

If you tried to build a row of houses that could accommodate all the building materials available for house exteriors, you would probably find yourself building on into infinity, for the range of materials and their possible combinations—in terms of shape, texture, and color—is virtually limitless.

Traditionally speaking, of course, wood, asphalt, and tile are commonly used for rooftop protection against the weather, and wood and metal for walls that aren't built of masonry. But there are also homes with roofs or walls of ribbed aluminum, plastic, foam, canvas . . . even sod.

In the following pages you'll find a generous collection of homes that show off traditional building materials to handsome advantage—and a few that make use of materials you don't often see. Let these ideas stimulate your creativity as you evaluate the personality of your own house exterior.

Staples of Modern Construction
House of wood (far left), roof of asphalt (top right), and walls of aluminum (bottom right) show how three staples of residential construction can be used for strikingly different effects in house exteriors.

POPULAR, DURABLE ASPHALT SHINGLES

Good looks, long life, and low maintenance make asphalt a favorite of the roofing class.

Asphalt Sampler
Cover your roof in any color or texture you like. Asphalt shingles come in frosty grays, bright greens and reds, and subtle earth blends, with either heavily textured or smooth-as-sand surfaces.

Clean Sweep
With their smooth, uniform surfaces, standard asphalt shingles reinforce the sweeping lines of a classic shed roof. Color variations add to the roof's visual appeal.

Distinctly Dimensional
Premium asphalt shingles cast deep
shadow lines to give a roof a distinct,
three-dimensional look.

Premium Patterns
Seen from a distance, heavy-weight
premium shingles such as these can
produce a strong diagonal pattern—
or a thoroughly random one.

Asphalt Shingles **7**

WAYS TO PUT EARTH TONES TO WORK

Warm, natural colors for asphalt roofs blend easily into any setting.

House of Several Gables
The pale fawn color of these shingles reflects heat from a multigabled ranchhouse.

Appealing Mix of Color and Cut
Smoky gray asphalt shingles appeal to the eye with random color shifts and irregular, staggered cuts along the bottom edge of each shingle.

In the Image of Slate
Handsome barnhouse exterior blends
slate-colored shingles with bleach-white
siding and mustard trim.

Restoring Charm to a Grand Victorian
Two-toned asphalt shingles conform handily to the cone-shaped turret of a restored
Victorian. The random grain in these standard shingles gives the roof its texture.

Deft Black Lines on Red Brick
Shingles in unbroken strips suit the clean
lines of a pine-clad contemporary.

Asphalt Shingles   **9**

THE MIXED MEDIA OF ASPHALT ROOFS

Draw from asphalt's wide range in color and texture to complement or contrast with your house's style.

From the Colonial Color Wheel
Green roof and shutters provide a pleasing contrast with white clapboard siding on a colonial-style home.

In the Dutch Vernacular
Two roofs—one a gable, the other a gambrel—dominate the architecture of a Dutch colonial. Here, a harmonious blend of grays in shingles, shutters, and house walls is underscored with crisp, white trim.

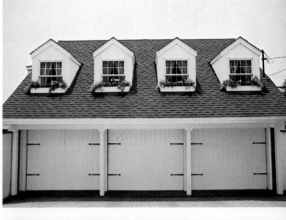

Quartet of Dormers
A strong, tailored feeling emerges from the combination of symmetrical black shingles with a quartet of dormers and window gardens.

Asphalt Shingles . . . or Wood?
Gambrel roof displays asphalt shingles that—because of their thickness and random widths—capture the warmth and style of wood.

Quiet Elegance
Dappled brown shingles, weathered spruce siding, and painted shutters breed a look of handsome sophistication in a home whose tall windows overlook a spacious garden.

ROOFS OF WOOD SHINGLES AND SHAKES

Warm to the eye and durable as well, wood roofing has a long suit in natural appeal.

Sampler of Shingles and Shakes
Thin wood shingles (top left) and their husky cousins, wood shakes (top right), are rich reddish brown when new, soft gray when weathered. Shingles known as "fancy cuts" (bottom) call attention to a roof even when used in modest proportions.

Serrated Shadows on a Wood Shake Roof
Because of their texture and the interplay of light and shadow, shake roofs are known for their bold, sculptural quality.

Plane Geometry, Saltbox-style
Weathered wood shingles, stone chimney, and white trim lend casual, comfortable charm to a saltbox-style home.

Shingled Shed
Wood shingles weave texture into a classic shed roof with an almost unbroken surface.

Snow Country Shakes
Hand-split shakes give this country cabin a woodsy, rugged look.

Random Widths for a Random Finish
Much of the textural quality in wood roofs comes from the random widths of the materials.

Wood Shingles & Shakes **13**

WOOD ROOFS WITH A CUSTOM LOOK

One is stained, one is staggered, and another is simply a work of art.

When Shingles Are Steamed
Over the rolling eaves, shingles were steamed into pliancy, then nailed in position.

Ocean Waves in Wood
Superb craftsmanship is behind the illusion of undulating "waves" that flow over an otherwise level roof surface. Varying the exposure in each ribbon of wood produced the mosaic-like pattern—with a result that's both eye-catching and restful.

Doubling Up
Four rows of doubled-up shingles
strengthen the horizontal pattern in a roof
stained black.

Decorate a Mansard with Diamonds
Diamond-shaped shingles—and rich green trim—call attention to the wall-like
mansard on an updated Victorian. Here's a nice balance of color and pattern.

Staggered Shingles, Smooth Siding
Against smooth siding, staggered shingles
on a mansard roof seem like ruffles.

ROOFING WITH CLAY AND CONCRETE

Roofing tiles, made of either clay or concrete, can last as long as the building they shelter.

Palette in Tile
Ribbed, barred, barreled, or flattened, roofing tiles are nearly impervious to the effects of weather. Red is the color for classic clay tile, though anything goes for concrete.

At Home with Wood
Red ribbed tiles on a cluster of shed roofs produce a netlike pattern. Here, tile is used both to offset and complement the fine horizontal lines and color of the cedar walls.

Far East Refrain
Borrowing its color from the Oriental pagoda, gently sloping blue tile roof combines ornamentation with utility in a wood and brick home.

Concrete Chips off the Wood Block
Imitating thick wood shakes, concrete tiles make a handsome counterpart to redwood siding on a contemporary ranch home. This house evolved from a tiny cottage whose frame required reinforcement before tiles could be laid. The weight of a tile roof was taken into consideration, though, when the additions were planned.

Barrels of Clay
Barrel-shaped clay tiles, indigenous to Mediterranean architecture, are right at home on Spanish-style residences.

TILES CREATED FOR SPECIAL EFFECTS

The idea of clay tiles for roofing came to us from the Mediterranean; the idea of ceramic tiles, from Europe and the Far East.

Pyramid in Emerald Green
Emerald green ceramic tiles add to the impact of a dramatic, angular roof line.

Quilt of Tiles
Ceramic tiles in shades of blue, laid randomly on a steep gable roof, create an interesting patchwork-quilt pattern for a townhouse facade.

Sensational Sculpture
Leading edge on a roof of gun-metal gray tiles has a sleek, sculptured curve. These tiles almost look poured into place, so smooth is the roof surface.

Gracious Tradition
Spanish-style home has a roof of clay tiles that have been painted black for a look of graceful elegance. These tiles have been in place since the house was built in the 1920s.

ROOFS OF METAL, SLATE, AND SOD

Five ways both traditional and modern materials merge with their surroundings.

Sod for a Solar Home
Soil held in place by a carpet of grass covers the north slope of a solar-heated home, helping to stabilize temperatures inside the house.

Tradition in Slate, Brick, and Stone
Slate, among the most traditional of all building materials, shingles the roof of a stately manor house built of brick and stone.

Unexpected Impact with Ice Blue Aluminum
In tandem with knotty cedar siding, this roof of ice blue aluminum makes an unexpectedly refreshing composition for a home exposed to a hot, dry climate. Fireproof and heat-reflective, the roof was installed in panels on either side of a ridge-length skylight.

Shingles of Aluminum
With their furrowed texture, muted color, and random thickness, aluminum shingles imitate the look of hand-split wood shakes.

Galvanized Steel Panels
Panels of mineral-coated galvanized steel, lightweight and easy to work with, take after clay and concrete roofing tiles.

Metal, Slate & Sod Roofs   **21**

HOUSES WITH WALLS OF WOOD

For variations in color, texture, pattern, and grain, no siding material is quite as versatile as wood.

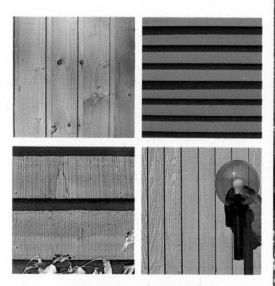

Four Ideas for Wood Walls
Inverted batten siding (top left), painted horizontal lap siding (top right), unfinished resawn lumber (bottom left), and stained plywood panels (bottom right) suggest the enormous range in wood species, widths, patterns, and finishes available for use on house exteriors.

The Natural Warmth of (Almost) Unfinished Wood
Protected simply by a water repellent finish, broad planks of knotty cedar on exterior walls and matching window casings give this ranch-style home a look of rustic charm.

Smooth Verticals
Smooth-surface tongue-and-groove siding, finished with a light-bodied stain, conforms to the sleek contour of a curve in the house wall.

Knot-speckled Pine
Speckled with knots and striated with random graining, smooth-sawn white pine was nailed horizontally over battens.

Board-and-Batten
You can strengthen the vertical lines of a house with classic board-and-batten walls.

SIDING WITH BOARDS AND PANELS

Look at board lumber, plywood panels, and sheets of hardboard when you consider clothing your house walls in wood.

Walls of Tweed
The "tweedy" look was achieved by sandblasting grooved plywood panels.

The Recycled Look of Rough-sawn Redwood
Rough-sawn redwood on a plywood base was left unfinished to underscore the wood's naturally rugged appearance

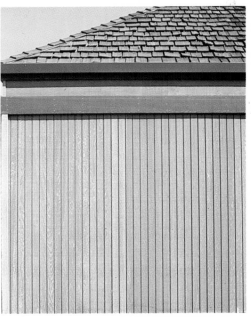

Plywood Siding with a Stucco Finish
Stucco walls go on quickly when they come in economical panels of plywood embossed with a stucco texture. Embossed plywood panels can simulate other surfaces as well.

Exterior Plywood
Plywood panels milled with grooves have the finished look of narrow board siding but go on more rapidly.

Study in Frank, Bold Design
Handsome, straightforward design incorporates triple lap siding stained to contrast with house trim.

INVENTIVE WAYS WITH WOOD SHINGLES

Use wood shingles to give your home as formal or eclectic a feeling as the house design permits.

Stock of Wood Shingles
Dressing a house with wood shingles—either standard cuts in random widths (top right) or decorative shingles cut to order (top left and bottom)—can give it as formal or eclectic a feeling as its basic design permits.

Shingled Geometry
Shingled walls and roof amplify the simple geometrics of a house that splits interior living areas into four different levels. Outdoors, the shingle-clad shell blends harmoniously with its wooded setting. A touch of color in the stovepipe adds just enough surprise to this pleasingly comfortable house exterior.

Jigsaw Treasures
A jigsaw and a little imagination — plus considerable patience — produced these delightful free-form shingles for an alpine retreat. A saber saw would also have worked.

Wall of Wood Feathers
Random widths and lengths of shingles give a feathery quality to house walls.

Fish-scale Finesse
Fish-scale shingles hug the curve of a cylindrical tower, emphasizing the shape of the lighthouselike annex and providing a pleasing visual contrast to the vertical board siding on the rest of the building.

Restored Queen Anne
With a fresh coat of paint, this restored Queen Anne shows shingles at their decorative best.

EASY CARE EXTERIORS OF ALUMINUM AND VINYL

Low maintenance, easy installation, and moderate costs make a persuasive case in practicality for aluminum and vinyl siding.

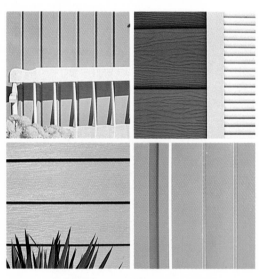

Finishes for Aluminum and Vinyl
You can color and texture your house any way you like with aluminum or vinyl siding. Vinyl, for example, can have a smooth, satin finish (top left) or woodgrain texture (bottom left). Walls of aluminum (top and bottom right) can resemble traditional lap or board-and-batten siding.

Cheery Walls of Canary Yellow
There's a cheerful, warm feeling about a three-story house designed with bold, uncomplicated lines and clad with bright yellow horizontal siding.

Barrel House of Blue Aluminum
Exterior walls and roof are one in a boldly designed, arch-roofed house that's nestled into a steep wooded setting. Smooth-finish aluminum panels were installed in a traditional manner over insulated walls. On the roof they were installed over a waterproof shell made of insulating polyurethane foam. Rain gutters the same color as the siding maintain continuity where the vertical walls end and the arched roof begins.

SIDING WITH ALUMINUM AND VINYL

Here are five conventional homes—both new and remodeled—refreshed with siding of vinyl or aluminum.

Aluminum Shakes for Side Walls
House walls were surfaced with 1 by 4-foot panels of aluminum "shakes" manufactured to resemble the sculptural quality of wood.

Contemporary Materials on a Classic Cape Cod
Crisp black-on-white color scheme captures the tailored look of tradition with modern building materials—the walls are re-sided with smooth-finish aluminum, the roof covered with asphalt shingles.

Barn Red Roof and Walls
Exterior walls of barn red aluminum
contrast handsomely with dappled red
roof and bright white window trim.

Colonial Charm
Newly constructed colonial blends gray vinyl siding with black-shuttered windows and
white trim for a look of quiet elegance. Close-up view of this siding shows its slightly
textured grain, which complements the smooth finish on the shutters.

Aluminum Board-and-Batten
Vertical aluminum siding has the look of
painted board-and-batten because of its
ribbed and raised surfaces.

Aluminum & Vinyl Siding **31**

ROOFING

If you have to go outside to remember the color of your roof, you have plenty of company—most of us rarely think twice about what goes on over our heads, until something happens to catch our attention.

Taken for granted though the roof often is, we depend on it to keep us safe and dry from wind and rain, and to protect the rest of our house and its contents. And on an esthetic level, we expect it to add to the house's architectural appeal.

Unfortunately, regardless of their quality, roofs eventually require repair or replacement. Asphalt, wood, and tile—the materials most roofs are made of—all suffer from their constant exposure to the extremes of sun, wind, and water. And when they fail, the very rains they were meant to keep out send us suddenly running for something to catch those offending drips.

But worse than turning our houses into badly tuned musical instruments, leaks, when ignored, will also produce dry rot in the sheathing and rafters—a structural problem that's difficult and costly to correct.

Since your home is likely to be your most important investment, it pays in both value and security to have a roof that offers proper protection. For you, it may simply be a matter of learning to inspect, repair, and maintain what essentially is a sound roof with many good years of life still in it.

But if on inspection you discover a roof that looks more like an oversized relief map than a uniformly shingled surface, there's a strong chance it's time to reroof.

Whichever direction you go—repairing or reroofing (or roofing a new structure)—let yourself be guided by information in the following pages. There you'll find instructions for conducting your own roof inspection, choosing among roofing materials, making standard roofing and gutter repairs, and—if you decide to do the work yourself—installing or replacing a moderately sloped asphalt, wood, or tile roof.

You'll also find information to help you decide whether or not to engage a professional in your project, and tips for selecting a roofing contractor if that's your decision.

And on page 36 there is a glossary of roofing terms to guide you as you use this book and also as you consult with suppliers and contractors.

A word of caution, though—before you scramble up to the rooftop, read *carefully* through the section on roofing safety (pages 40–41). Roofing repairs and replacement generally aren't complicated jobs (it's more a matter of "sweat labor" in roof work), but you must exercise extreme caution when you work at rooftop heights, on slippery or uneven surfaces, around power lines, and in the awkward postures that roofing tasks often require.

ROOFING STYLES

The next time you go for a walk or a jog, look for the six roof styles commonly used in residential architecture (see illustration at right). Very likely you can find them all—the gable, gambrel, and hip roof popularized in Europe in the 17th cen-

Six views of the basic roof

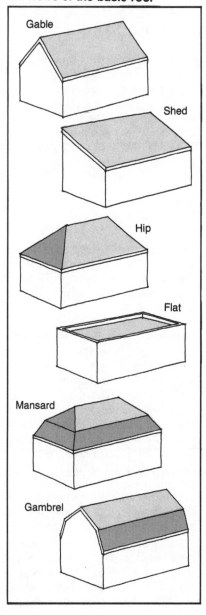

Gable

Shed

Hip

Flat

Mansard

Gambrel

Asphalt Shingles
Handsome, durable, and easy to apply, asphalt shingles—both standard and premium—make a popular choice for new roofs and reroofing.

tury, the wall-like mansard roof, and shed and flat roofs that have been in use for thousands of years. You'll probably see not only these, but a number of their variations and combinations as well—perhaps a gable and shed roof, or a gambrel and a shed, or a hip roof modified to make room for dormer windows.

Of course, simple shed roofs with straight edges and no "valleys" (see "Terms the Roofer Uses," page 36) are the easiest roofs to repair and install. The more complex the roof style and the steeper the slope (pitch), the more difficult the job becomes.

ROOF STRUCTURE

Typically, a roof is put together pretty much like the floor or walls of your house. It has a framework of rafters to support a roof deck (sometimes called a "subroof"), which consists of sheathing and underlayment; the roof deck, in turn, provides a nailing base for the roof surface (see illustration).

The roof deck

The type of roof deck a house has depends primarily on the nature of the roof surface material, but most decks have both sheathing and underlayment.

Sheathing. Sheathing is the material that provides the nailing base for the roof surface. Sloping roofs with asphalt shingles usually have solid plywood sheathing—4 by 8-foot panels nailed directly to the rafters (though in older homes you may find 1 by 6 or 1 by 8-inch boards). Homes with open-beam construction may reveal fiberboard sheathing that has been manufactured so one side forms the nailing base for the roof, the other the finished ceiling of the house.

Roofs of thin wood shingles or thick wood shakes (see page 44) are often laid over "open" sheathing— 1 by 4-inch lumber spaced evenly over the rafters to permit air circu-

The skin and bones of a roof

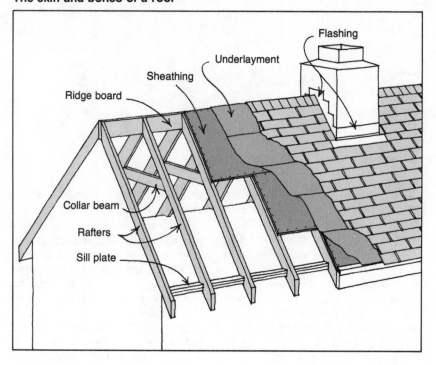

lation. In some instances, particularly where wood shingles or shakes are put down over an existing roof, you'll see horizontal batten boards nailed directly onto the old roof surface.

Underlayment. Sandwiched between the sheathing and the roof surface is the underlayment, usually roofing felt. Roofing felt is a thick, fibrous black paper made from wood chips and recycled paper, then saturated with asphalt. It is thick enough to resist water penetration from the outside, thin enough to allow moisture from within the house to escape.

To produce a more fire-resistant roof system, some wood shake roofs have underlayments of both roofing felt and metal sheeting. Where extra waterproofing is necessary, such as in hurricane zones of the Southeast, tile roofs often have underlayments made with built-up layers of roofing felt and hot-mopped asphalt.

The surface

Most sloping roofs are covered with overlapping layers of asphalt shin-

gles, wood shingles or shakes, or tile, though the possibilities range from sod at one extreme (see page 20) to molded plastics at the other.

Flat—built-up—roofs are most frequently covered with alternating layers of roofing felt and asphalt, with a layer of gravel on top. In some cases, though, they're surfaced with an insulating polyurethane foam that's sprayed on and painted with a protective coating.

The principle underlying the function of the shingle roof is simple: to shed a drop of water that falls on a sloping roof by drawing the drop gradually down the side, over layer upon layer of lapped material, until it falls to the ground.

Flashing

Wherever water is likely to collect and penetrate the roof surface—along the joints around a chimney, at the edges of roof vents, or in the "valleys" where two roof planes meet at an angle—at all those places, protective flashing is necessary.

Made of malleable metal or plastic, flashing appears as the drip edge you see along the eaves of a roof, the collars around ventilation pipes, and the "steps" along the chimney. Less obvious flashing also protects the points where solar panels and television antennas are connected to the roof.

Roof slope

The slope, or pitch, of a roof refers to the vertical rise measured against a standard horizontal distance of 12 inches (see illustration at right).

The term "4 in 12," applied to a roof, tells you that the roof rises vertically 4 inches for every 12 horizontal inches. Very low-sloped roofs measure only 1 in 12 or 2 in 12; steeply sloped roofs range from 12 in 12 (a 45-degree angle) up to 20 in 12.

Ordinarily, home craftspersons should be able to work safely on roofs with slopes no steeper than 6 in 12.

Determining roof slope. To find out the slope of your roof, all you need is a carpenter's level, a metal tape measure, and a flat piece of wood. Working either on the roof or from a ladder at the house eave, form a right triangle with the level, board, and tape measure; the tape and level should intersect 12 inches from the point where level and board meet (see illustration). Then measure the vertical distance from the board sur-

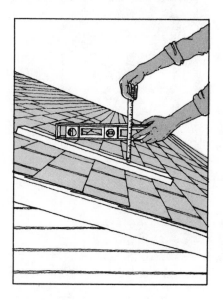

Roof slope

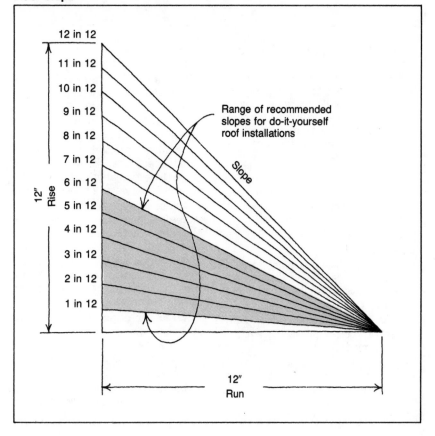

Range of recommended slopes for do-it-yourself roof installations

12 in 12
11 in 12
10 in 12
9 in 12
8 in 12
7 in 12
6 in 12
5 in 12
4 in 12
3 in 12
2 in 12
1 in 12

12" Rise

Slope

12" Run

face to the bottom of the level. The number of inches, followed by "in 12," gives you the slope.

REPAIR OR REROOF? THE TWO-STEP INSPECTION

Roofs are a little like shrubs. Many of us have met the overgrown shrub in shears-to-branch combat and wondered how in blazes we could make this a thing of beauty (again). And most of us have at least once heaved a sigh over the results of our labors and wished we had called in a professional, since clearly the shrub had won.

So it is with roofs. Periodically inspect and maintain yours and you will have a serviceable, easy-to-maintain roof. Ignore it and one day you may be on the phone with a roofing contractor, talking about replacing dry-rotted sheathing or water-soaked roofing insulation or deteriorated underlayment.

Learning to do your own roof inspection isn't at all difficult, and it's satisfying: the more you know about any part of your house, the better equipped you are to maintain it yourself.

The first step to take in a roof inspection is to learn the age of the roof and its expected life span (see "Buyer's Guide to Roofing Materials," pages 48–49). (If possible, ask the previous homeowner when the roof was put on.) Then, in light of your climate, evaluate how much longer you can expect it to last. If, say, you live in a moderate climate and you have a 15-year-old asphalt shingle roof that's supposed to last 20 or 25 years, you can expect to find a somewhat worn but still serviceable roof. On the other hand, if yours is a 25-year-old wood shingle roof with a 20-year life expectancy in a moderate climate, you're likely to find a roof that needs replacing.

Begin your inspection from inside the attic (or the house, if you have

Words like "exposure," "rake," and "valley" that are bandied about freely in the roofing industry probably don't have quite the same meaning for you. Looking at them through a roofer's eyes, though, will streamline the project you undertake. Here are some of the more commonly used terms:

Built-up roof: A flat or slightly sloped roof surfaced with alternating (or built up) layers of roofing felt and hot-mopped asphalt, and a layer of gravel or crushed rock on top.

Butt: The exposed end of a shingle or shake.

Courses: The horizontal rows of roofing material, laid successively from the eave to the ridge of the roof.

Deck: The structural nailing base for the roof surface, usually composed of wood or plywood sheathing and felt underlayment.

Drip edge: A type of flashing made of thin strips of metal or plastic that extend the length of eaves and rakes to facilitate water runoff.

Eave: The edge of the roof that projects beyond the house wall.

Exposure: The portion of each shingle or tile exposed to the weather. Also called "weather exposure."

Fascia: Wood or other trim covering the ends of the rafters.

Flashing: Waterproofing materials, usually metal, that connect roof shingles or tiles to chimneys, valleys, vent pipes, vertical walls, eaves, and rakes.

Gable: A type of roof with two slopes meeting at a horizontal ridge. Also, the triangular area formed by such a roof.

Hip: A downward-sloping intersection of two roof planes that extends from the ridge to the outside corner of the house. Also, a type of roof.

Rake: The edge of a pitched roof at the gable end.

Ridge: The top edge of the roof, where two roof slopes meet in a horizontal line.

Sheathing: Boards or plywood sheets that form the nailing base for roofing shingles or tiles.

Slope (or pitch): The number of inches of vertical rise of the roof over a horizontal distance of 12 inches; a "4 in 12" roof has a slope that rises 4 inches over a 12-inch run.

Soffit: The underside of the rafters and roof at the eaves.

Square: Unit of measure equaling 100 square feet, used as a basis for measuring roof area. Also, the amount of roofing material, allowing for overlapping, needed to cover 100 square feet of roof.

Tab: The cutout part of an asphalt shingle; three-tab shingles usually have three 5-inch tabs in a strip that measures 12 by 36 inches.

Underlayment: The material, usually asphalt-saturated roofing felt, used to cover deck sheathing before the roof surface is put down.

Valley: The junction where two downward sloping roofs meet at an angle; an important channel for water runoff.

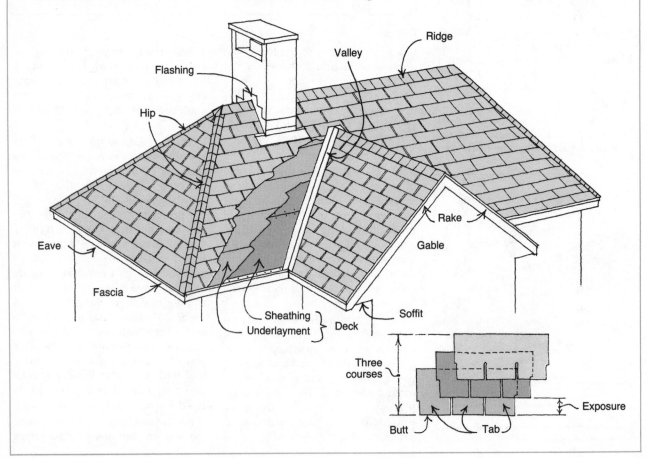

open-beam ceilings); then head outdoors for the exterior check. Tools you'll need are a good strong flashlight, a thin screwdriver, a scraping knife, and a ladder (see page 41).

When to make a roof inspection

It's a good idea to inspect your roof in autumn, just before the hard weather hits (and when you're likely to be up there anyway, clearing out the leaf collection in gutters and valleys). Then in the spring, when you have an opportunity to assess whatever damage winter may have done, inspect the roof again.

The view from the attic: The interior check

When you're looking at your roof from the attic or crawl space, your primary concern is evidence of prolonged water contact with wood rafters and sheathing.

Starting with the ridge where rafters are joined, poke with your knife and screwdriver to locate any soft spots in the wood, especially where you notice dark water stains. Dried water marks are signs of leaks, either at their source or along their paths down rafters.

If you have a wood shingle roof, don't be alarmed if you see small shafts of light coming in at an angle; when wet, the shingles swell small cracks shut. But if you see holes directly above you, stuff wire or drive nails through them so they'll show up well when you make a closer examination from the roof. It may be that the wood has become "sunburned" (worn through) and needs to be replaced.

In an attic insulated with fiberglass, in the form of blankets or rigid foam board, pull back sections in areas where you suspect moisture. (When working with insulation, wear protective clothing, gloves, goggles, and a protective mask.)

Finally, while you're up there, take a look at the amount of ventilation provided. A surprisingly large amount of water damage to roof rafters and sheathing can come from within the house. Moisture from cooking, bathing, plant respiration, human respiration and perspiration, washers, dryers, and dishwashers must be controlled by vapor barriers or carried off by ventilators before the moisture settles into rafters and sheathing. Proper ventilation also minimizes heat buildup.

Though you should check your local building codes for recommended amounts of ventilation for your own home, as a rule of thumb you can estimate 1 square foot of ventilation for each 150 square feet of attic floor area. If the attic has vapor barriers, 1 square foot of ventilation for each 300 square feet of attic floor space may be adequate. (See page 63 for additional information.)

The view from outdoors: The exterior check

When you examine the roof from outdoors, evaluate the condition of the roof structure, the surface material, flashings, eaves, and gutters.

Before you head up the ladder, though, thoroughly read the section on roofing safety (pages 40–41). And if you're at all uneasy about being up on the roof, if your roof is higher than two stories, or if it is steeply sloped (see "Roof Slope," page 35), then use a pair of binoculars and make your check from a ladder or from the ground.

Check roof structure. First, standing back away from the house, look at the lines of the ridge and rafters. The ridge line should be perfectly horizontal. The line of the rafters—which you can assess by looking along the plane of each section of the roof—should be perfectly straight unless the roof was intentionally designed with a curve.

If ridge or rafters appear to sag a little like Aunt Effie's swayback horse, call in a professional contractor. It's possible you have a structural problem only professional skills can remedy.

Inspect the roof surface. Next, either from the ladder or on the roof, inspect the roof surface for signs of wear, loose or broken nails, and missing shingles.

If you have a sloped roof, check the south side first. Though roofs tend to wear evenly, the south side will wear first because it absorbs the full force of the sun's rays.

Asphalt that's aging may show bald spots where the mineral granules have worn away (if the roof is very old, you'll probably find granules accumulating in the gutters and at the feet of downspouts); large bare patches indicate the roof is beyond its serviceable age. Look also for curled shingles, missing shingles the wind may have torn away, and cracks in shingle tabs and in ridge and hip shingles.

Next, pinch off a small corner from one or two shingles. If the core appears black, fine; the shingles still have protective oils in them. But if the shingles appear gray and bloated and if the material crumbles easily between your fingers, it's probably time for a new roof—the shingles have become wicks, rather than sheds, for water.

Finally, bend back the ends of several shingles and press them with your fingernail or a screwdriver. Those with life in them will be flexible and resilient; those without will be brittle and easily broken. Since brittle shingles are easily torn off the roof by wind, they should be replaced.

Wood shingles that have curled with age or been worn thin and brittle by the sun are also vulnerable to winds, which will snap them off. If yours is a wood shingle roof, look for curled, broken, and split shingles, and for spots where nails have become loose or rusted. Look for signs of "sunburn," too, especially at the spots you plugged with wire or nails during your attic inspection.

The extent of the defects you find should indicate whether repairs will suffice or whether replacement is necessary. (If your wire-plugged roof reminds you of a bristle brush, you probably need to reroof.)

Wood shakes show their age when the wood crumbles easily underfoot or between your fingers. Thick and durable though they are, shakes also wear thin under constant exposure to water, heat, and light, so you'll want to look for areas where shakes have worn through to expose the sheathing. Also look for split and loose pieces where nails have rusted through.

Tile or slate that's used as a roofing material isn't likely to wear out, though individual tiles do chip and break and require replacement. To check the felt underlayment, which may give out and need to be replaced, remove tiles from several locations on the roof. If the underlayment is black and resilient to the touch, it's still in good condition. If it appears gray and bloated and crumbles easily, it probably needs to be replaced. In that case, the tile or slate must be removed temporarily and then put back in place.

Asphalt ("tar") and gravel, or built-up roofs, are perhaps the most difficult to diagnose, but you can look for these indications of wear: bare patches where the gravel has blown away, "rosettes" of asphalt that have blistered above the gravel, a separation between the roof surface and drip edge, curling or split roofing felt that's exposed, and cracks in the surface along the north and east roof edges.

Flashings, eaves, and gutters. In valleys and in flashings around the chimney and vents, watch for broken seals along the flashings' edges and for rust spots in the metal. Leaks often begin here, and recaulking may be all that's necessary to check them. Flashings that appear to have rusted through in several places most likely need replacement.

Use your knife and screwdriver to test the boards along roof eaves. Since most leaks flow toward the eaves, water damage is most commonly seen here. Often you can scrape out any damage caused by wood rot and fill the holes with a caulking compound (see "Correcting for moisture and dry rot," page 77).

If you have metal gutters and downspouts, look for rust spots that have either weakened the metal or worn holes through it (see "Inspecting Gutters," page 81). Plastic gutters generally need only periodic cleaning to keep them in working order, since they are invulnerable to both wood rot and rust.

WHEN REPAIRS ARE ENOUGH

If your roof inspection turns up nothing more than a few missing or split shingles, defective flashings, or rain gutters clogged with debris, the necessary repairs can probably be made rather simply, and you will probably choose to make them yourself. Where you run into rotted sheathing, eaves extensively damaged by water, or underlayment that needs replacing, you may want to call in a roofing contractor.

Doing your own repairs

To make your own roof repairs, all you really need are the proper tools (see pages 42–43) and materials, a little dexterity, and adequate time to do the work properly.

Even if you've never been on a roof before, with proper caution you should be able to replace missing shingles or broken tiles, renail loose shingles, or repair gutters and flashings. Replacing small areas of rotted sheathing in the attic, or splicing a damaged rafter with a "sister" rafter may also be well within range of your skills and experience.

You'll find instructions for making standard roofing repairs on pages 72–77, "Making Roof Repairs."

Where a contractor can help

If you're looking at a roof that has suffered extensive damage, or one whose slope is threateningly steep, or if you simply aren't interested in doing your own work, you're wise to engage a contractor.

He or she probably has experience in correcting structural problems, repairing damage done to large areas of sheathing and underlayment, and replacing sizable sections of surface materials damaged by wind, rain, or falling branches.

Choose a contractor as you would if you were going to have the whole roof replaced. Investigate. Get bids from several different contractors, and ask for several references covering work they've done that's at least 2 years old. Then, after you've followed up on the references and chosen a contractor, see that a detailed contract is prepared and signed.

WHEN THE ROOF SAYS "REPLACE ME!"

If your inspection has turned up an asphalt roof with wide-open bare spots where the mineral granules have worn away, or a wood roof whose shingles have snapped like twigs or crumbled like compost, it's time to reroof. And if that's the case, the next question is whether to do the work yourself or have it done professionally.

Roofing on your own

There's hardly a home improvement project that, successfully done, doesn't offer both tangible and intangible rewards—and roofing is no exception. Aside from the pleasure that comes from knowing you did the work yourself—and to your own liking—you are also rewarded by the chunk of money saved in labor costs.

The rising cost of labor, in fact, is one reason more and more homeowners are doing their own maintenance and repairs. Labor eats about half the pie in a typical roofing project; eliminate that and you're on your way to bisecting the cost of your roofing project.

Unfortunately, while the idea of a new roof at half price is pretty appealing, you must remember that you will be the one to do the work. And roofing is characteristically a hot, tiring, and tediously repetitive job with some sizable hazards built into it. You must decide realistically whether or not you are willing and physically able to see the project through to completion.

Do you qualify for the job? If you want to do your own work, you should have some handiness with tools, patience in following instructions to the letter, scrupulous regard for safety procedures (see pages 40–41), and a willingness to adapt basic procedures to the eccentricities of your own house.

You should also be in good physical condition. Kneeling over a slope all day, nailing shingles in their courses, will tax both your endurance and your backside, not to mention the arm and shoulder that must drive all those nails. And you'll be scooting up and down the ladder and working at roof heights; if you're at all bothered by heights, you're better off turning the project over to a professional.

How suppliers can help. The next time you pass a home improvement or building supply center, drop in and pick up some of the brochures (most are free) put out by roofing manufacturers. They illustrate the abundant range in roofing materials and offer helpful tips for roofing installation and repair. You may also find sure-fire roofing kits, accompanied by complete instructions, that some manufacturers are packaging for the do-it-yourself roofer.

When you need a professional hand

Not all roofing jobs belong to the home craftsperson. You're probably over your head, so to speak, if you're thinking about doing your own work on

- a steeply pitched roof (more than 6 in 12 —see page 35)
- an unusually large surface area
- a house with multiple roofs or more than two stories high

- a roof that requires additional structural bracing
- a roof with numerous dormer windows or other architectural detours away from simple, broad planes.

And, though it's possible to do your own work, you're also likely to strain your skills if you plan to roof with tile. It is heavy, hard to cut (you'll need a carborundum blade on a circular saw), difficult to handle, and slippery underfoot. If you're reroofing with tile, you may be required by code to have a structural engineer evaluate the load-bearing capacity of your house's framework. If the framework can't easily support the extra weight, you'll need to call in a contractor to beef it up.

Flat roofs covered either with layers of roofing felt and hot-mopped asphalt or with polyurethane foam are also beyond the skills of most homeowners, because of the processes and equipment involved. Working with boiling hot asphalt is a dangerous, difficult, and messy business.

Contracting the work out

If you decide to turn your roofing project over to a contractor, take as much care in choosing one as you would for any other remodeling project. Get bids and client references from several roofers, then see that a detailed contract is drawn up with the roofer you select.

Choosing a contractor. The best recommendation a roofer has is past work. In fact, the roofer depends on client referrals for almost 80 percent of his or her jobs. So feel free to ask for names and addresses of several clients whose roofs have been installed 2 years or longer.

If friends or neighbors can't recommend roofers they've worked with, look under "Roofing Contractors" in the Yellow Pages of your telephone directory. You also may be able to get recommendations from a local branch of the Roofing Contractors Association (RCA). Select several roofers who indicate they specialize in the type of roof you want; then call and ask for bids and references.

Because bids often double as contracts, they should include everything that's pertinent to the project: the type of materials to be used, tasks the roofer will perform, an indication of when the work will be done, and financial arrangements and guarantees.

When you review the bids, compare them for the types of materials specified and the cost of labor. If you're interested in streamlining costs, ask each contractor how much you can save if you tear off the old roof, or do your own gutter replacement, or haul debris from the job.

When you check references, also call the Better Business Bureau and ask if any complaints have been filed against any of the roofing companies you're considering. Check, too, with the county clerk to see if any lawsuits are pending.

Finally, find out how long the company has been in business. Those that have been in the area for some years will most likely be around to correct a problem that may occur a few years down the road.

The contract. "If it's worth saying, it's worth writing down" is the tune to whistle as you review the contract you plan to sign for a roofing project.

The contract should spell out very clearly every aspect of the project, from brand names of materials to method of payment.

Be sure the contract identifies, by brand name and weight, the type of roofing material to be installed. Just "asphalt shingle," for example, isn't good enough; you may have in mind a premium quality, heavyweight shingle, and your contractor may install something quite different. If underlayment, flashings, gutters, and drip edges are included in the price, they too should be listed in the contract.

Tasks you expect the roofer to perform should be itemized. Does the contract price include preparation and repair of the roof deck? Replacement of fascia or gutters? Removal and reinstallation of solar panels? Removal of debris from the site? Or will these be done only for an additional time-and-materials charge?

See that guarantees are spelled out. Most roofers guarantee their work for 2 or 3 years, so make sure this and other promises—such as replacement of landscaping that could be damaged during the progress of the work—have the roofer's signature behind them.

The contract should also specify a reasonable schedule for completing the work—usually within 30 days after the contract has been signed.

Licensed contractors are required by law to carry worker's compensation insurance. For your own protection, get it in writing that your contractor has it.

Identify a payment plan. Some roofers ask for a down payment and then a lump sum when the work is finished. Never pay the full amount until the job is done to your satisfaction. If the roofer has hired subcontractors, get it in writing that those subcontractors have been paid, so you'll have protection against liens.

A WORD ABOUT BUILDING CODES

However removed your own house may be from the clustered hubs of urban housing, it's still subject to rules designed to protect you from faulty construction practices. Just about the last thing you want is a new tile roof that drops in unex-

pectedly on your dinner party because no one told you the rafters weren't strong enough to support it.

So check with your city or county building official to learn which codes may affect your roofing project. Typically, the codes will specify

• numbers of reroofs permitted (usually two) on the same structure

• types of materials that must be used—for example, in fire-hazardous areas or in historic areas or other places where esthetic standards must be maintained

• structural requirements for roofs surfaced with tile or slate

• appropriate types of sheathing for different roof materials.

Your inspector can also tell you whether or not a permit is required for your project. The cost of the per-

WORKING SAFELY WHEN YOU'RE OVER YOUR HEAD

Before you embark on a trip up your roofing ladder, it's important that you know and observe the following safety precautions and use, where necessary, proper safety equipment.

General considerations
Keep these tips in mind every time you approach the roof.

• Wear loose, comfortable clothing, rubber-soled shoes with good ankle support, and a hat for sun protection.

• Work on the roof only in dry, calm, warm weather. A ladder or roof that's wet from rain, frost, or dew can be treacherously slick, and a sudden wind can knock you off balance. Keep your shoes clear of wet grass and mud as well. Never get on the roof when lightning threatens.

• Once on the roof, be alert for slippery, brittle, or old roofing materials, and rotten decking you could put a foot through.

• Avoid contact both with power lines connected to the roof and with energized television antennas.

• To avoid straining your back, lift

only lightweight loads, and let your leg muscles do the work.

• Pace yourself and take frequent rests. As soon as you feel fatigued, stop work for the day.

• Keep children and pets away from the work area; they can be hurt by falling materials.

Safety equipment
Several of the standard safety devices illustrated are available from tool rental companies for homeowners who prefer not to buy them.

• Metal ladder brackets allow you to hook a ladder over the ridge of a house.

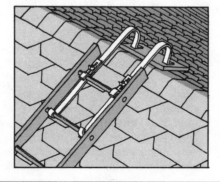

• Roof brackets or toe board jacks, when nailed to the roof, support yourself and your materials on a 2 by 6 plank. (Use strong, straight-grained lumber no longer than 10 feet unless you support the middle with another bracket.) The brackets have notches in them so they can be slipped off the nails when layers of shingles have been applied above. (Secure brackets and jacks with long nails that will penetrate wood decking or rafters; locate nails where materials will later cover them.)

mit may be a flat rate or a percentage of the total project cost.

INSURANCE

Before the new roof goes on, check into your medical and liability coverage. Later, with a new roof that increases your property value, you may want to upgrade your homeowner's policy.

Medical coverage

Unfortunately, in spite of cautions, people do fall off roofs, legs break when ladders fail, thumbs turn purple and other colors under the blow of a hammer . . . and up on the roof, most of these accidents happen to amateurs. So make sure you're covered. Check your insurance policies to be certain they provide proper medical coverage for you, liability coverage for friends who volunteer their help, and worker's compensation for laborers you may hire.

Worker's compensation insurance. Though its provisions vary from state to state, worker's compensation insurance generally covers costs of treatment—surgical, medical, and hospital—in case of an occupational injury. It also reimburses the injured worker for wages lost.

You can get a worker's compensation policy through your insurance broker, directly from an insurance company, or through a state fund if your state has one.

Registering as an employer. If you employ people directly and if they will earn more than the minimum amount set by the state, you must not only have a worker's compensation policy, but you must also register with the state and federal government as an employer, withhold and remit income taxes and disability insurance, and withhold, remit, and contribute to social security.

Upgrading your policy

Applying a new roof adds value to your property, so you may want to increase the value of your insurance policy when the new roof is installed. On some policies you can get a rate reduction if you've re-roofed with fire-resistant materials.

● An angled seat board allows you to sit on a level surface while working. Angles on the sides of the board must match the slope of your roof.

● A safety belt or harness attached with a halyard to a ¾-inch nylon throw rope allows you to work safely on steeply sloped roofs. The rope is fastened to something sturdy, such as a toe board jack nailed upside down on the opposite side of the roof, or the trunk of a tree on the opposite side of the house. A series of loops knotted in the rope will enable you to work safely at different levels.

● Scaffolding can be rented from tool supply companies.

Ladders

Inspect your ladder for cracks or weaknesses in the rungs before you lean it against the house. Then place the ladder on firm, level ground at a measured distance from the side of the house—that distance should equal a quarter of the vertical distance from the ground to the top rung.

If the ladder is to stand on a slick surface, install rubber safety shoes (they're available at home improvement centers).

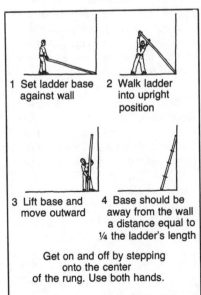

1 Set ladder base against wall

2 Walk ladder into upright position

3 Lift base and move outward

4 Base should be away from the wall a distance equal to ¼ the ladder's length

Get on and off by stepping onto the center of the rung. Use both hands.

ROOFING TOOLS AND MATERIALS

Even after 5,000 years, it seems that in terms of roofing materials we still don't have much on the Babylonians—for asphalt, our most popular roofing material, is the same stuff our eastern ancestors used for waterproofing their own buildings.

Technology being what it is, of course, no transported Babylonian would recognize asphalt roofs the way they come to us today—packaged in tidy bundles of shingles complete with installation instructions and warranties.

And where the Babylonian was probably less concerned with esthetics than with hard-core practicality, today's roofers consider visual impact as well as the reason why the roof is there in the first place.

Perhaps the concern for esthetic appeal has grown because so many roofing materials—asphalt, wood, and tile, as well as slate, aluminum and metal—have been made available to us on so grand a scale.

If you're having a new roof installed, use the following pages to learn about the virtues and drawbacks of different roofing materials. The more you know about them, the better equipped you will be to make a choice that suits both your budget and the architectural style of your house.

If you're planning to do the work yourself, you'll also learn that it's not necessary to haul your entire home workshop up to the roof; surprisingly few tools are necessary to get the work done.

TOOLS FOR ROOFING

Aside from safety equipment and a good ladder (see pages 40–41), the range of tools required for roofing is relatively modest (see illustration on next page). The tools most commonly used include these:

- roofer's hatchet for nailing and aligning wood shingles in their courses, and for splitting and scoring wood shingles or shakes (for asphalt roofs you can use an ordinary claw hammer with a smooth rather than waffle head);

- utility knife, with extra blades, and a straightedge for cutting asphalt shingles;

- 24-inch wrecking bar for prying out nails and lifting shingles;

- tin snips to cut metal flashings;

- 25-foot metal tape measure to plot vertical and horizontal alignments;

- string line and blue or yellow chalk to mark guidelines for shingles;

- steel trowel or putty knife to apply plastic roofing cement (and solvent to clean the tools);

- plastic roofing cement and caulking gun;

- wire brush to prepare areas for flashings and patching;

- tool belt with loops for small tools and a pocket (or nail stripper) for nails;

- gloves to protect your hands when working with metal flashings;

- circular power saw with a carborundum blade, to make precise cuts in tile; and

- square-bottomed garden spade with a short handle, to tear off an old roof.

Be prepared to add other tools to your inventory as the need arises—such as a hack saw for cutting nails whose heads are hidden.

ROOF DECK MATERIALS

What materials are used in the sheathing and underlayment of a roof often depends on the type of surface that's going down.

Sheathing

Most roofing materials are laid over solid sheathing, though wood shingles and shakes are often laid over spaced lumber. This "open" sheathing system allows air to circulate freely around the roofing materials.

Types of solid sheathing. Plywood is most commonly used for roof decks that require solid sheathing, though 2 by 6 board lumber, manufactured deck panels—even the old roof surface—also are used.

Plywood is generally favored because it comes in 4 by 8-foot panels that are easy to install over rafters (see "Installing plywood decks," pages 52–53). It's also strong and durable, and it doesn't easily warp under changing weather conditions.

Wood boards measuring 1 by 6 or 2 by 6 are often milled with tongue-and-groove edges that interlock for more stability (see "Installing tongue-and-groove boards," pages 52-53).

Homes with exposed beam ceilings may have decks of tongue-

Tools the roofer uses

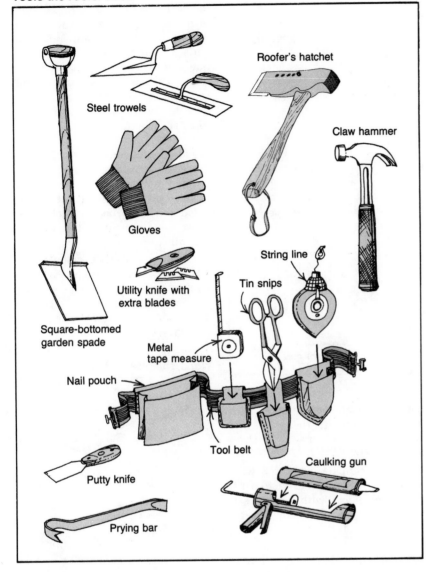

Roofer's hatchet

Steel trowels

Claw hammer

Gloves

Utility knife with extra blades

String line

Tin snips

Square-bottomed garden spade

Metal tape measure

Nail pouch

Tool belt

Caulking gun

Putty knife

Prying bar

and-groove lumber, or manufactured fiberboard that combines a solid nailing base for the roof surface, thermal insulation, and a finished underside for an attractive ceiling.

Spaced or "open" sheathing. Wood shingles are commonly laid on a deck of spaced, rather than solid board, sheathing. The boards, usually 1 by 4s, are laid horizontally, several inches apart, across the rafters or over the existing roof (see page 53).

Choosing wood for sheathing. If you're using wood to replace old

sheathing or to install a new roof deck on a room addition, select plywood or lumber that is dry and well seasoned; a roof laid over "green" lumber (lumber with more than 20 percent moisture content) may buckle when the sheathing twists and warps as it dries.

If you plan to put down plywood sheathing, check local building codes to learn the permissible thickness for your rafter span. Though many codes permit the use of plywood as thin as $5/16$ inch on roofs with 12-inch rafter spans ($3/8$-inch thicknesses on 24-inch spans), you may want to go with something thicker ($1/2$ or $3/4$ inch) for a sturdier

nailing base. Plywood used on roofs should be made with exterior grade glues but need not have exterior type veneers.

If you're installing board sheathing, use well-seasoned tongue-and-groove planks no wider than 6 inches (wider boards will warp more easily). Eliminate those with numerous knots or pockets of resin that could penetrate and damage the roof surface.

Underlayment

Roofing felt is the water-resistant membrane sandwiched between the roof's sheathing and surface. Made from wood fibers and recycled paper that have been saturated with asphalt oils, roofing felt prevents rain from penetrating and damaging the sheathing. It also prevents the roof surface from coming in contact with moisture or resins in the wood.

Roofing felt is milled in various thicknesses and identified according to the approximate weight—15 pounds, 30 pounds, and 90 pounds—per square of roof surface. It's generally sold in 36-inch-wide, 36-foot-long rolls which, allowing for overlapping, cover about one square of roof.

Asphalt roofs usually have underlayments of 15-pound felt; wood shake roofs have 30-pound felt underlayments laid in overlapping courses. Roll roofing and valley flashing on some asphalt shingle roofs are made from 90-pound felt.

SURFACE MATERIALS

If you're putting down a new roof, you face many decisions. In the sea of roofing materials, wading from one shore to the other—and coming through with a decision in hand—can be an unexpectedly foggy experience. But it can also be fun, and certainly rewarding.

First you must decide among the types of materials that are available: asphalt, wood, and tile, or the less-frequently used aluminum shingles, galvanized steel panels, and slate. Then if you plan to roof with asphalt shingles, you have to decide between those manufactured with an organic base and those manufactured with a fiberglass base. Finally, you must choose among a prodigious assortment of colors, textures, and patterns—of which some will blend nicely with your house siding, and some won't.

Types of roofing

Asphalt, wood, and tile are the three materials most commonly used in residential roofing today, though you can find roofs covered with slate, aluminum or galvanized steel . . . even plastic or sod. Flat or very low-sloping roofs usually have surfaces of either tar and gravel (actually asphalt and gravel) or a polyurethane foam that provides both insulation and weatherproofing.

Asphalt shingles. Asphalt shingles are a popular choice for residential roofing because they are economical, attractive, widely available, easy to install, and easy to maintain. In their manufacture, mats made from either wood pulp and paper fibers or fiberglass are covered top and bottom with protective layers of asphalt and then coated on top with mineral granules of various colors.

Organic-base asphalt shingles have felt mats made of wood and paper fibers and are commonly known as "asphalt" or "comp" (short for composition) shingles.

Fiberglass-base asphalt shingles, or just "fiberglass" shingles, have a fiberglass mat as the shingle base.

Both are available by the square from roofing supply companies. Standard three-tab shingles (see above, right) measure 12 by 36 inches, though some manufacturers are producing slightly larger metric-size shingles. They measure .336 meters (13¼ inches) by 1 meter (39⅜ inches).

Most asphalt shingles sold today have built-in protection against wind

Asphalt shingles

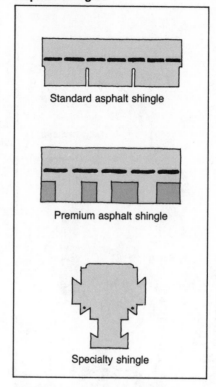

Standard asphalt shingle

Premium asphalt shingle

Specialty shingle

because they are manufactured with a self-sealing mastic that welds one shingle tab to another after the shingles are installed.

Asphalt shingles come in a wide variety of colors, from rich earth tones that blend with natural settings to crisp whites, reds, greens, and blacks.

Shapes range from traditional rectangular butts that give a roof a smooth, clean appearance, to random-edge butts that provide a more rustic look. Premium-weight shingles (300, rather than the standard 220 to 235, pounds per square) are manufactured with random-width tabs and additional layers of asphalt for a three-dimensional look comparable to that of wood shingles.

Specialty shingles—such as 9 by 12-inch ridge shingles—are also available. In some areas you may find wind-resistant interlocking (called "T-lock") shingles, 12 by 18-inch "Dutch lap" shingles, or hexagonal shingles that come in strips or rolls, though most of these are being phased out in favor of the self-sealing three-tab shingle.

Asphalt roll roofing. Asphalt roofing is also manufactured in the form of roll roofing, 36 inches wide and 36 feet long, with a mineral grain surface. Roll roofing is popular on outbuildings and economy housing with slopes as low as 1 in 12. Roll roofing of a matching color may be used to cover the valleys of an asphalt shingle roof.

Wood shingles and shakes. Despite the relatively modest number of roofs covered with wood in the United States—only about 10 percent—wood has often been imitated because of its natural beauty and durability.

Shingles, with their smooth, finished appearance, are sawn from chunks (called "bolts") of western red cedar. They come in lengths of 16, 18, and 24 inches. Most shingles and shakes are nailed one by one to

Wood shingles and shakes

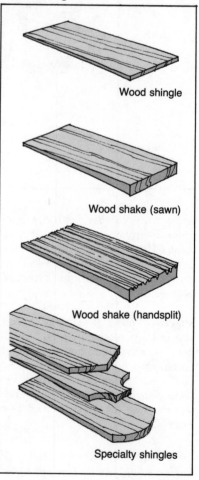

Wood shingle

Wood shake (sawn)

Wood shake (handsplit)

Specialty shingles

the roof, though you can find pre-nailed materials in 8-foot panels for faster application.

The thicker shakes are split by machine or by hand from the bolt into 18 and 24-inch lengths. You can buy them in two weights, medium and heavy, and in several styles, such as handsplit, handsplit and resawn, and taper-split.

Handsplit shakes generally last longer than machine-split shakes, because the wood fibers haven't been sawn through and are thus less inclined to rot.

Though shingles and shakes are available in several grades (they are used in siding as well as roofing) be sure to specify Number 1 grade if you're planning to roof with either one. This grade requires that the wood be all heartwood (the most durable, sap-free part of the tree) and straight grained.

Clay and concrete tiles. Until recently, classic red clay roofing tiles—kiln-dried in the shape of a barrel (see illustration at right) and made to last the life of a building—were an exclusive highlight of Mediterranean and Spanish-style homes of the West, the Southwest, and Florida. They are a familiar feature of California's earliest architecture, the Spanish missions.

With the introduction of equally durable concrete tiles—made from a blend of cement, sand, and water—the use of roofing tiles has flourished in recent years.

Because they are extruded, concrete tiles can be manufactured in shapes that are quite unlike the Spanish barrel. You'll find concrete tiles that are flat, ribbed, S-shaped, even textured to resemble wood shakes.

Typically, tiles measure 12 by 17 inches and are ½ inch thick. Lugs on the bottom of concrete tiles hook over furring strips that are nailed to solid decking. (Clay tiles are fastened with nails or wires.) Manufacturers also produce tiles with interlocking edges, and accessory ridge, hip, and rake tiles.

There's a whole spectrum of colors

Clay and concrete tile

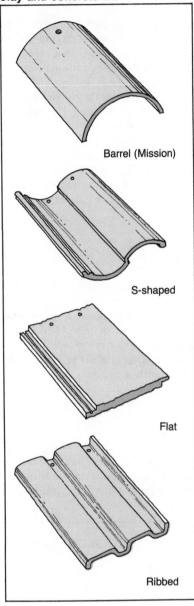

Barrel (Mission)

S-shaped

Flat

Ribbed

in concrete tiles, with subtle earth tones at one end and shades of blue, yellow, red, green, and black at the other. With so many choices of style and color, roofing tiles have become increasingly popular in contemporary as well as traditional residential design.

Weight is the most critical factor if you're considering roofing with tile. Tiles weigh 900 to 1,000 pounds per square—three to four times as much as asphalt shingles—and your roof must be sturdy enough to handle the weight.

Check with your building department to see if you'll need an engineering report. Your house may require additional structural bracing.

You may also want to evaluate shipping costs if you live a considerable distance from manufacturers. Tile is more expensive than most other roofing materials, and particularly expensive to ship because of its weight; you may find that the combined costs are beyond your budget.

Finally, consider carefully the hazards involved if you want to install a tile roof yourself (see "When you need a professional hand," page 39).

Slate and aluminum shingles. In some parts of the country houses may have roofs of slate, aluminum, or plastic tiles.

Slate, a cousin to tile, is either mined from quarries in Vermont and Virginia or imported. Though slate offers unmatched permanence, it's generally used only in custom roofing because its cost—$275 to $350 per square—is beyond the range of most home improvement budgets. And it usually requires professional installation.

Aluminum shingles, shaped to interlock with each other for additional wind resistance, are sometimes preferred in areas of heavy snow because of their durability. Often corrugated, they come in various colors.

Metal panels. Panels of galvanized steel are favored as roofing in some situations because they are easy to install, can be painted any color, and shed snow easily. Some manufacturers produce metal panels that resemble tile.

Built-up roofing systems. Homes with flat or low-sloping roofs may have surfaces of either asphalt and gravel or polyurethane foam.

Asphalt (or "tar") and gravel roofs are made with several layers of roofing felt, each coated with hot or cold-mopped asphalt. The uppermost layer is then surfaced with crushed rock or gravel.

Polyurethane foam is sprayed on new roof decking or on existing built-up roofs. Though usually a little more expensive than asphalt and gravel roofs, polyurethane foam roofs have the advantage of providing both durable roofing and insulation in a lightweight material.

Warranties

Most roofing manufacturers guarantee their products against manufacturing defects during the expected life of the roof, but only on condition that the roof has been put on correctly. To validate the warranty on roofing materials you purchase, be sure to stick closely to the printed instructions that come with the product—using the prescribed number of nails, for example, and allowing the recommended weather exposure.

Most warranties are also prorated. If, for instance, you have a 20-year warranty on $1,000 in materials, your warranty is worth the full $1,000 if the roof fails in the first year. After 10 years, though, the warranty is worth only $500.

Fire ratings

Underwriters Laboratory, an independent testing service, has tested all types of roofing materials to determine their resistance to igniting, supporting the spread of fire, and adding to a fire hazard by emitting burning brands.

Materials are rated Class A (for materials with the most fire-retardant qualities), Class B, and Class C. Because of their flammable nature, some materials such as untreated wood shingles or shakes receive no rating.

HARDWARE & MATERIALS

When you plan for roofing materials, don't overlook flashings, roofing cement, and nails.

Flashings

Flashings for valleys, chimneys, drip edges, and vents are most commonly made of a malleable, 28-gauge galvanized sheet metal, though on asphalt shingle roofs, valleys and vent pipes can also be flashed with mineral surface roll roofing. Plastics and aluminum may be used too, and copper is sometimes preferred for chimney flashing. You can either buy preformed flashings for drip edges, valleys, and vents, or you can make your own (see "Flashings," pages 57–62).

Plastic roofing cement

Plastic roofing cement is a black, elastic waterproofing agent used along flashings, under shingles, and between layers of felt. Available in caulking tubes or 1 and 5-gallon cans, roofing cement is usually applied with a caulking gun or steel trowel.

Nails

The type and size of nails to be used in roofing depends on the nature of the project. Generally, hot-dipped

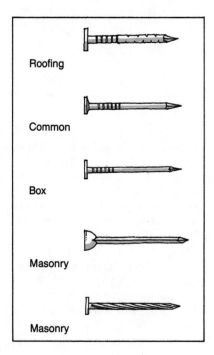

Roofing

Common

Box

Masonry

Masonry

galvanized nails are recommended for underlayments and roof surfaces because these nails won't rust through and damage surrounding materials. They should be long enough to penetrate ¾ inch into the roof deck, or all the way through plywood sheathing. Less expensive common or box nails may be used when applying sheathing.

With the introduction of the pneumatic stapling gun, applying shingles has become a faster-moving activity. But unless your warranty covers application with staples and you are familiar with the use of the equipment, you're wise to stick to using nails. Applied improperly, staples may not fasten shingles securely to the deck. They can also break through the asphalt.

TIPS FOR CHOOSING ROOFING

The slope of your roof (see page 35) and local building codes will eliminate rather quickly the types of materials that aren't suitable for your project. After that, learn all you can about the remaining materials and compare them for:

• cost versus durability (premium-quality asphalt shingles, for example, may be less expensive than standard asphalt shingles on a cost-per-year basis);

• warranties (some provide only material replacement, others also cover labor);

• ease of application, especially on steeply sloping roofs;

• availability of materials (shipping charges may price you out of the market); and

• appearance.

When you compare roofing materials for appearance, look at your house as a whole and try to blend the roofing with the type of siding you now have or plan to have later on.

Browse through the color photographs on pages 4–32 (as well as those in manufacturers' brochures) and observe how roofing colors harmonize with exterior siding. As you choose your own roof color, your goal is to strike a similar balance. In your research you'll find that, on the whole, the most appealing house exteriors make use of no more than two or three colors. You'll also notice that:

• the weathered colors of wood shingles and earth tones of asphalt shingles usually blend well with natural settings;

• bright colors such as reds, greens, and blues emphasize a roof;

• neutral, medium tones will play it down;

• whites and light colors will make the house seem higher, and will reflect heat;

• dark colors will make the house seem more compact and will absorb heat.

ESTIMATING & ORDERING MATERIALS

Estimating for roofing materials is a fairly simple matter of finding the number of squares (one square = 100 square feet) in the roof surface.

How many squares in your roof?

If you have a simple roof with unbroken planes, finding the number of squares in it is a fairly simple process. First figure the surface area of each of the roof's rectangles by multiplying the length by the width (see illustration below). Next, total

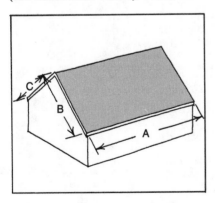

the sums and divide by 100, rounding off to the next highest figure.

For more complex structures and steeper roofs, start by measuring the square footage of the house at ground level. Then, based on the slope of the roof (see "Determining roof slope," page 35), use the tables below to compute the roof area.

Slope	Multiply by
2 in 12	1.02
3 in 12	1.03
4 in 12	1.06
5 in 12	1.08
6 in 12	1.12
7 in 12	1.16
8 in 12	1.20

Here's an example: Suppose you have a house and a garage like the ones illustrated below. The house measures 30 by 40 feet (1,200 square feet) at ground level, and the roof has a 4 in 12 slope. The garage measures 10 by 30 feet (300 square feet), and its roof has a 6 in 12 slope.

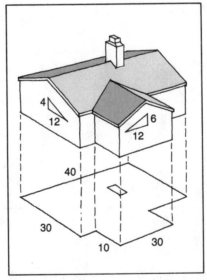

To compute the roof area over the house, multiply 1,200 by 1.06, according to the table above. The roof area here measures 1,272 square feet.

Then, to compute the roof area of the garage, multiply 300 by 1.12 for a total of 336 square feet.

The total roof area is 1,272 plus 336, or 1,608 square feet; divided by

100, the roof area computes to a little more than 16 squares.

When you estimate your own roof surface, adjust for dormer windows, skylights, chimneys, overlapping roof planes, and overhangs at the eaves.

Estimating

Include in your estimate your requirements for roofing shingles, ridge and hip shingles (and rake shingles if you're roofing with tile), lengths of flashing for the drip edge, roofing felt, and nails.

Estimating for shingles and felt. To the number of squares of shingles and felt you'll need, add 10 percent to your estimate to allow for waste, double coverage of shingles or tiles at the eaves, and future small repairs.

How to estimate hip and ridge shingles. With asphalt shingles, you can make 100 feet of hip and ridge shingles (12-inch squares with 5-inch exposures) from each square of roofing shingles.

To estimate hip and ridge shingles for other materials , measure the lengths of hips and ridges. Then, to learn how many shingles you need, divide the total by the exposure recommended for the shingles.

How to estimate flashings. Measure the total lengths of eaves and rakes for drip edge flashings, and the lengths of valleys for valley flashing. Vent pipe flashing is sold by the unit. Measure the outside dimensions of chimneys and determine the chimney flashing pattern (see pages 59–60) before you order materials.

Estimating nails. The number and type of nails you'll need will depend on the roofing material you plan to apply; however, you can figure roughly 2½ pounds of nails per square of asphalt shingles, and 2 pounds per square of wood shingles or shakes. You'll also need about 2 pounds of nails for the starter course and hip and ridge shingles.

When you order

Call several roofing suppliers to compare material costs and delivery charges before you order.

To save yourself the backbreaking labor of loading materials onto the roof yourself, *have materials delivered onto the roof*—especially if you're putting down shingles or shakes over open sheathing. Most roofing suppliers have hydraulically operated scissors trucks specially designed for rooftop deliveries.

If you expect a considerable lapse of time between purchase and application, try to arrange to have materials delivered when it's time to install them. (It's a good idea to have some plastic sheeting on hand to protect exposed sheathing or underlayment, if you foresee a delay in delivery of shingles.)

Storing materials. If you must have materials on the site for an extended period, store them indoors in a dry

BUYER'S GUIDE TO ROOFING MATERIALS

		Cost Per Square		Weight Per Square (Pounds)	Durability	Fire Rating
		Materials	Labor			
Sloped Roofs	Asphalt shingles (Felt base)	$35–40 (Standard) $40–75 (Premium)	$30–40	240–300	12–20 years depending on sun's intensity	C
	Asphalt shingles (Fiberglass base)	$35–40 (Standard) $40–75 (Premium)	$30–40	220–300	15–25 years, depending on sun's intensity	A
	Wood shingles and shakes	$60–90	$40–50	144–350	15–25 years, depending on slope, heat, humidity	None, untreated; C, when treated with fire retardant; B, with use of fire retardant and foil underlayment
	Tile (concrete and clay)	$50–80	$50–60	900–1000	50+ years	A
	Slate	$200–350	$50–100	900–1000	50+ years	A
	Aluminum shingles	$120–130	$80–100	50	50+ years	C or better
Flat & Low-Slope Roofs	Metal panels (Aluminum or steel)	$50–200	$50–200	45–75	20+ years	C or better
	Asphalt roll roofing	$15–20	$20–30	90–180	5–15 years, depending on water runoff at low slope	C or A
	Asphalt ("tar") and gravel	$30–40	$80–100	250–650	10–20 years, depending on sun's intensity	C
	Sprayed polyurethane foam	$80–100	$80–100	20 for 1-inch thickness	Life of building with proper maintenance	A

place, and away from extreme temperatures.

If you must leave them outdoors, stack materials off the ground on 2 by 4's and cover them with plastic to protect them from the rain.

Loading materials. If for some reason you must load your own materials onto the roof, either rent a mechanical hoist from a tool supply company or raise materials using a rope strung through a pulley rigged to a ladder. Don't try to carry materials up the ladder a bundle at a time—it's not only time-consuming, but hazardous as well.

You can find pulley systems at roofing supply companies. Have a helper work from the ground to slip-knot the rope over each bundle and send it up the ladder. When you have materials on the roof, scatter the bundles along the ridge to distribute the weight properly. (Tiles should be distributed evenly around the roof in stacks of six.)

Recommended Minimum Slope	Merits	Drawbacks
4 in 12 and up; down to 2 in 12 with additional underlayment	Available in wide range of colors, textures; easy to apply and repair; conforms easily to curves in roof surface; low in maintenance; economical.	Less durable and less fire-resistant, though equal in cost to fiberglass-base shingles.
4 in 12 and up; down to 2 in 12 with additional underlayment	Durable and highly fire-resistant; available in wide range of colors, textures; easy to apply and repair; conforms to curved surfaces; low maintenance.	Brittle when applied in temperatures below 50°F/10°C.
4 in 12 and up; down to 3 in 12 with additional underlayment	Appealing natural appearance with strong shadow lines; durable.	Flammable unless treated with fire retardants; treated wood expensive; time-consuming application.
4 in 12 and up; down to 3 in 12 with additional underlayment	Extremely durable; fireproof; available in flat, curved and ribbed shapes; moderate range of colors.	Costly to ship; difficult to install; requires sufficient framing to support weight; cracks easily when walked on.
4 in 12 and up	Attractive, traditional appearance; impervious to deterioration; fireproof; available in several colors.	Expensive and costly to ship; difficult to install; requires sufficient framing to support weight; may become brittle with age.
4 in 12 and up	Lightweight; fire-resistant; made to resemble wood shakes; moderate range of colors.	Can be scratched or dented by heavy hail, falling tree branches.
1 in 12 and up	Aluminum: lightweight; durable; maintenance-free for prepainted panels; sheds snow easily. Steel: strong; durable; fire-resistant; sheds snow easily.	Contraction and expansion of metal can cause leaks at nail holes; noisy in rain.
1 in 12 and up	Economical, easy to apply.	Drab appearance.
0 in 12 and up	Most waterproof of all roofing materials.	Must be professionally applied; difficult to locate leaks; black surfaces absorb heat.
0 in 12 and up	Continuous membrane produces watertight surface; good insulation value; lightweight; durable when protective coating is maintained.	Must be professionally applied; quality depends on skill of applicator; deteriorates under sunlight if not properly coated.

HOW TO PREPARE THE ROOF DECK

Whether you're roofing over old shingles or new construction, you'll be rewarded if you use care when you prepare the deck, since both the appearance and the durability of the finished roof depend largely on the quality of the deck.

When you prepare the deck, take care of other roof changes that should be made before the new roof goes down. If, for example, you want to add a skylight over the kitchen, install it now (see "Cutting through the Roof," pages 78–79). Perhaps you plan to install a ridge ventilation system; it too should be in before shingles are applied (see "Roof Ventilation," page 63). If yours will be a tile roof, avoid having to replace broken tiles later by having the television antenna installed before tiles are put down.

Should you intend to install a solar hot water system, on the other hand, wait until after you reroof—solar panels belong on top of the roof.

Some of the smaller jobs—such as repairing or installing gutters, or putting on new drip edges—also should be coordinated with roof deckwork. New drip edges, for example, should be installed along the eaves before the underlayment is applied, and along the rakes afterward (see page 61).

ROOF OVER . . . OR TEAR OFF?

If you're reroofing, your first job is to decide whether you can roof over the old surface or should tear it off and begin with a stripped deck.

Using the existing roof as the deck for the new surface is undoubtedly the easiest route to a new roof—it saves you the time and trouble of stripping the old shingles off, and it provides an extra layer of insulation where the house loses heat the fastest.

But whether you can roof over the old surface—or whether you're better off stripping away worn materials down to the sheathing (or rafters)—depends on several factors:

- the condition of the present surface and sheathing;

- the compatibility of old and new surface materials;

- the number of roofs the framework already supports;

- the manufacturer's recommendations for the material you've chosen; and

- local building codes and requirements.

When you can roof over

Building codes and manufacturers' warranties permitting, most asphalt and many wood shingle roofs will serve as an adequate deck for new materials.

Roofing over asphalt. If your old roof is asphalt, you generally can roof over with asphalt shingles, wood shingles or shakes, or, if the rafters and house framing are strong enough, tile.

Be sure, though, that both the old surface and the sheathing underneath are in good condition. If the sheathing has been badly damaged by water, it will be necessary to remove both the old roof and the damaged sheathing. (Small areas usually can be patched.)

The asphalt shingles should lie flat—curled shingles underneath will give your new roof a lumpy look. If you find spots with a few missing shingles, you can replace them before roofing over; but if you find large bare areas, it's a good idea to strip the roof down to the sheathing.

If you're roofing with wood—especially wood shingles—you should install furring strips (see page 54) for ventilation between the two roof surfaces.

Roofing over wood shingles. A wood shingle roof will usually provide a nailing base for new roofs of wood, asphalt, and in some cases tile.

Wood shakes will last longer if they're put down over spaced sheathing, but you can install them over a wood shingle roof if you sandwich a layer of 18-inch-wide, 30-pound felt between courses of shakes.

Wood shingles will also go down on an old wood shingle roof, though many professional roofers advise against this: a new wood shingle roof will last longer, they believe, when not in contact with older, possibly rotted wood. For a more durable roof, you may prefer to rip off the old shingles and install the new shingles over spaced sheathing so they can breathe.

Asphalt shingles also are compatible with wood shingles. If the wood roof does not lie flat, a new layer of 30-pound felt should be laid over the wood shingles.

Roofing over wood shakes. Where engineering reports determine that the roof can support the additional weight—and where building codes permit them—tile roofs have been installed over shake roofs in some communities. The tiles either rest on or are fastened to 1 by 2s that have been nailed to the butt ends of the shakes. Otherwise, shakes should be removed (see below).

When to tear the old roof off

You'll have to tear the old roof off if you're dealing with a badly worn surface, rotted sheathing that should be replaced, or the maximum number of roofs allowed by code; and in most cases if it's a wood shake roof.

Badly worn surfaces. You're better off stripping the surface down to the sheathing if you have a roof with extensive bare spots where shingles have blown away, or large areas where the shingles are badly curled, or, on roofs with wood shingles or shakes, considerable deterioration of the wood. These conditions spoil the uniform, stable nailing base you need for a new roof.

Rotted sheathing. If you've found large areas of water damage in the sheathing, plan to strip off the old roof and repair those areas before you apply the new materials (see "How to replace damaged sheathing," page 77).

Too many reroofs. If you already have the maximum number of reroofs allowed by your local building codes, you may be required to remove all of them before you put on a new surface.

Usually you're permitted three roofs—the original and two reroofs—on the same framework; with wood roofs, though, most codes allow only two. Your building inspector can tell you what the limits are in your community.

Wood shakes. Wood shake roofs must be ripped off before new asphalt shingles, wood shingles, or shakes are installed. The thick texture of wood shakes is far too irregular for most reroofing, which requires a flat surface.

Tile or slate. It's very unlikely you'll ever need to roof over tile or slate; these materials are so durable they generally outlast several underlayments. When they do need to be replaced, the old tiles must be removed entirely.

Incompatible sheathing. If you plan to replace a wood shingle roof with asphalt shingles, you may be required to remove the old roof and substitute solid sheathing for spaced sheathing in order to provide the new material with a sounder nailing base. Check with your building inspector.

PREPARATIONS FOR ROOFING OVER

To prepare the old surface for a new roof, see that all shingles lie flat, and replace any shingles that are missing. Nail down asphalt shingles that have curled or buckled; split and nail down curled or warped wood shingles. Strip off ridge and hip shingles, removing nails as you work, so new materials will lie smoothly. Then sweep the roof clean of debris.

To hide the old roof completely, use a hatchet (for wood) or a utility knife (for asphalt) to strip away 6 inches of roofing material from the eaves and rakes. Then nail on a 1 by 6-inch board and add a drip edge as described on page 61. Besides camouflaging the old roof, the board provides a firm base for nailing along what is often the weakest part of the old roof.

Next, check to see that all chimney and vent flashings are in good order; check, too, to see if mortar on the chimney needs repair.

If you're roofing over asphalt shingles, install new valleys made out of roll roofing or galvanized metal (see "Valley flashings," pages 57–59). If you're roofing over wood shingles, install new valleys of galvanized metal (see pages 57–58).

Do you intend to repair or replace gutters when you reroof? Then coordinate this work with your preparation of the deck; some gutter fasteners must be nailed to the deck before shingles are applied (see "Gutters and Downspouts," pages 80–84).

As a final step, measure roof surfaces carefully from eave to ridge and from rake to rake, to see if they are perfectly rectangular. If not, you'll know that you need to compensate when you align courses for the new roofing material.

Make compensations near the ridge by shortening the last few horizontal courses. To adjust vertical alignments, hide uneven shingles at the least obvious rake.

TEARING OFF THE OLD ROOF

Beginning at the ridge of the roof, use a square-bottomed garden spade with a short handle to lift and remove the old shingles, as the sketch below illustrates. (Have a refuse bin located next to the house eaves for the discarded materials. You can

probably rent one by the week from your garbage collection service.) If you have two or three roofs, remove all of them, one at a time.

Be careful not to damage any flashing; even if it has deteriorated, it's useful as a pattern for new flashing.

When the old shingles have been removed, pull nails and, if necessary, repair damaged sheathing (see "Repairs for sheathing and rafters," page 77) or install new as described below.

If you find areas with knots or resin pockets, clean out as much resin as possible with a knife or putty knife. (Clean the tool with solvent when the job is done.) Then cover the areas—and large cracks as well—with squares of 26-gauge galvanized metal nailed to the sheathing (see sketch below).

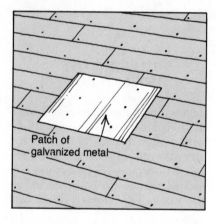

Patch of galvanized metal

NEW & STRIPPED DECKS

For both new construction and stripped roofs, prepare a deck that suits the type of roofing material you plan to apply over it.

Perhaps the most important thing to keep in mind as you install sheathing and underlayment is that roof surfaces may not measure out as perfect rectangles. Careful measuring before you begin will help you to fit sheathing and compensate for irregularities when you apply underlayment and shingles.

Decks for asphalt shingles

Asphalt shingles should be applied on a deck with sheathing of solid plywood or wood boards and an underlayment of 15-pound felt (see "Underlayment," page 43).

Installing plywood sheathing. When installing plywood sheathing, lay the plywood horizontally along the roof, with the grain running perpendicular to the rafters (see illustration on next page).

Stagger the panels so vertical seams aren't in line with one another. To allow for expansion, leave a 1/16-inch space between the ends of adjoining panels, 1/8 inch between sides. If your climate is exceptionally humid, double the spacing.

If you're putting down 3/8-inch plywood over 24-inch rafters, install H-shaped plywood sheathing clips along the sides, every 2 feet between rafter spans. They will ensure that surfaces lie flat. (Clips are available by the box for different thicknesses of plywood; buy them at lumber supply stores.)

When nailing plywood to the rafters, use 2-inch common or box nails for plywood up to 1/2 inch thick, 2 1/2-inch nails for plywood 5/8 to 7/8 inch thick. Drive a nail every 6 inches along the vertical edges of each panel, and every 12 inches where a panel crosses a rafter.

To adapt plywood to irregular shapes, carefully measure the area and transfer measurements to the plywood before cutting.

At the ridge, plywood edges should line up with the center of the ridge beam.

Once the plywood is laid, install drip edges at the eaves before applying underlayment (see page 61).

Installing tongue-and-groove board sheathing. Choose well-seasoned boards in lengths of 4-foot multiples so you can join them where they cross rafters. Tongue-and-groove joints should fit snugly together, but you should allow 1/8-inch expansion space where board ends meet. Use two 2 1/2-inch common or box nails

to fasten each board wherever it crosses a rafter, as the illustration on the next page indicates.

Applying underlayment. Before you roll out the roofing felt, correct defects in the wood surface by cleaning out resin pockets and knotholes with a knife or putty knife. Then nail squares of 26-gauge sheet metal over knots, resin pockets, and any large cracks in the surface. Also install the drip edge (see page 61) along the eaves.

To make the alignments even, measure the roof carefully and snap horizontal chalk lines before you begin. To snap a chalk line, hook one end of the line to one edge of the roof. Then, stretching it taut across the roof plane, lift the line a foot from the roof surface, and let it snap back. The chalk line will be your guide.

Snap the first horizontal chalk line 35 5/8 inches above the eave (this allows for a 3/8-inch overhang). Then, providing for a 2-inch overlap between strips of felt, snap each succeeding chalk line at 34 inches.

When applying felt, start at the eave, leaving a 3/8-inch overhang along the drip edge, and lay the strips horizontally along the roof, working in layers toward the ridge (see illustration on next page). Felt should be trimmed flush at the rake and overlapped 6 inches at the valleys, hips, and ridges. Where two strips meet in a vertical line, overlap the material 4 inches.

Drive just enough nails to hold the felt in place (about one 1 1/4-inch galvanized roofing nail for every square yard of felt). Near the eave, use shorter nails if the overhanging deck is thinner than the length of standard roofing nails.

Once the underlayment has been applied, install flashings as needed (see "Flashings," pages 57–62).

Decks for wood shakes

Wood shakes are usually laid over a deck of solid plywood or board sheathing, though in some instances

Installing plywood sheathing

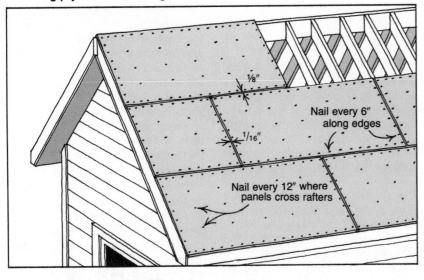

Nail every 6" along edges

Nail every 12" where panels cross rafters

1/8"

1/16"

Tongue-and-groove sheathing

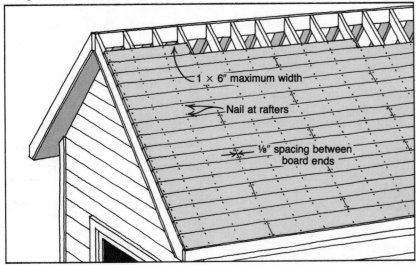

1 × 6" maximum width

Nail at rafters

1/8" spacing between board ends

Applying underlayment

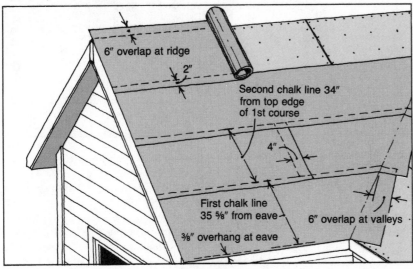

6" overlap at ridge

2"

Second chalk line 34" from top edge of 1st course

4"

First chalk line 35 5/8" from eave

6" overlap at valleys

3/8" overhang at eave

they are put down over spaced sheathing. Either way, strips of 18-inch-wide 30-pound felt are interlaid between courses of shakes.

Installing plywood or board sheathing. Follow instructions for "Decks for asphalt shingles" (preceding) when you put down either plywood or board sheathing, being sure to cover knotholes and resin pockets with small squares of sheet metal.

Installing spaced sheathing. To install spaced sheathing, lay well-seasoned 1 by 4-inch lumber horizontally along the roof, using another 1 by 4 as a spacing guide. Fasten each board to the rafters with two 2½-inch nails, allowing ⅛-inch spacing where boards meet (see sketch below).

⅛" spacing between board ends

3½"

Installing furring strips. If you're putting down wood shakes over solid sheathing or asphalt shingles, you may want to install furring strips. If so, follow instructions given below under "Decks for wood shingles."

Applying interlayment. The easiest way to apply interlayment is to place it after you nail each course of shakes. See page 70.

Decks for wood shingles

Wood shingles are usually laid over a deck of spaced sheathing—with no underlayment—so that the shingles are exposed to maximum air circulation. (Shingles, sawn smooth on both sides, lie flat and can trap water. Shakes, on the other hand, don't need the same air circulation; because they are split, they have rough grooves that form natural channels for carrying water off and allowing air to move between them.)

In some areas, building codes permit wood shingles to be applied over solid sheathing that's been covered with 15-pound felt. If you use wood shingles over solid sheathing or asphalt shingles, you must build up the surface with furring strips to allow for air circulation between the shingles and the roof.

Installing spaced sheathing. Follow preceding instructions for applying spaced sheathing for wood shakes. Then repair or install new flashings where necessary (see "Flashings," pages 57–62).

Furring strips for solid decks. When you apply furring strips over solid sheathing or old asphalt shingles, use well-seasoned 1 by 4s. Lay the first board flush with the eave line, as illustrated below, and from

there space boards so their centers are the same distance apart as the exposure for the shingle—5 inches for a 5-inch exposure, 7½ inches for a 7½-inch exposure, and so on.

Extend the boards ½ inch beyond the rake, and allow ⅛-inch spacing where boards meet. At the ridge, butt two strips together. At the valleys, leave ½ inch between abutting strips.

Decks for tile roofs

The usual deck for a tile roof has three parts: solid sheathing, an underlayment of either one layer of 30-pound felt or two layers of 15-pound felt, and—for some concrete tiles—1 by 2-inch furring strips spaced horizontally over the roof on top of the felt.

The furring strips accommodate the projecting anchor lugs of concrete tiles when they are laid over the strips. On roofs with 7 in 12 and steeper slopes (or those exposed to high winds), the tiles are nailed to the strips. Tiles without anchor lugs are usually nailed (or occasionally

wire-fastened) directly to the roof on top of the felts.

If you live in hurricane country in the southeast, your building codes may require a more watertight underlayment composed of several layers of asphalt felt with hot-mopped asphalt in between (a job for professional roofers). Tiles are then nailed securely to furring strips or mortared to the underlayment for extra protection against high winds and horizontally driven rain.

Other building codes may allow for concrete tiles to be nailed directly to the sheathing. In some cases they may also be laid over spaced boards—a practice not recommended because moisture seeping through a cracked tile can easily penetrate the sheathing and rafters.

Installing sheathing. Follow instructions described on pages 52–53 for installing solid sheathing for asphalt shingle roofs, with the following modification:

Instead of extending the plywood to the center of ridge and hip beams, lay the plywood just to the edge of the beams, as illustrated below. This

Furring strips: wood shingles and shakes

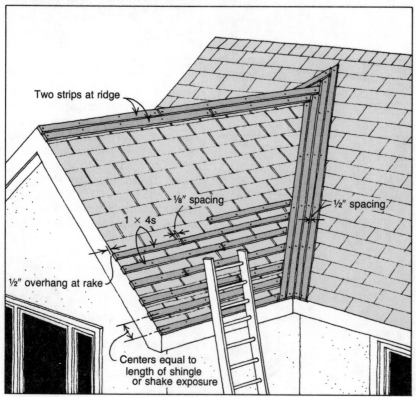

Two strips at ridge

⅛" spacing

1 × 4s

½" spacing

½" overhang at rake

Centers equal to length of shingle or shake exposure

Building up the ridge: tile roofs

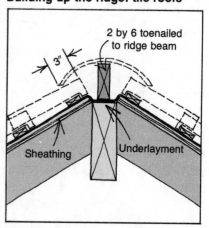

2 by 6 toenailed to ridge beam

3"

Sheathing

Underlayment

way, the ridges and hips can be built up—with 2 by 2s, 2 by 4s, or 2 by 6s—to accommodate ridge and hip tiles. Along the ridge, 2 by 4s (for flat tiles) are nailed flat to the beams, or 2 by 6s (for curved tiles) are set on edge and toenailed to the beams after underlayment is applied. Various dimensions of lumber are used to build up hips, depending on the shape of the tile and the manufacturer's recommendations.

Applying underlayment. Use either 30-pound felt or two layers of 15-pound felt for underlayment, extending the felt 6 to 12 inches across the ridge and hip beams to the other side of the roof. Otherwise, instructions are the same for applying underlayment to asphalt shingle roofs (see pages 52–53).

Installing furring strips. When tile roofs require them, install furring strips at intervals equal to the exposure of the tile (measured between the upper edges of successive strips).

First, if you're installing flat tiles, nail a furring strip flush with the eave, as the sketch at right indicates. ("Birdstop" flashing—see page 71—is nailed at the eave when curved tiles are installed.) Then lay a sample tile along the eave at each rake, allowing for the recommended overhang—usually 1¼ inches for a roof with gutters, 2 to 3 inches for one without.

Mark at both rakes where the lugs of the tile will grip the upper edge of the furring strip (from the top of the tile to the lug grip is often 1 inch). Then snap a chalk line between the two marks.

Next, install 4-foot-long sections of furring strips, placing their upper edges against the chalk line. Fasten them to the roof with 1½-inch nails wherever they cross a rafter.

To permit water runoff, allow ½-inch spacing between abutting strips, or place a 2-inch square of asphalt roofing between the strip and the felt wherever you nail.

Next, install a furring strip at the roof ridge, with its upper edge one inch (or the distance from the upper edge of the tile to the bottom of the lug) from the edge.

Then measure the distance between the upper edges of the two furring strips. Divide the distance by the recommended tile exposure. (Measure and calculate three times for accuracy.)

If your calculations yield a whole-number multiple of the recom-

Furring strips: tile

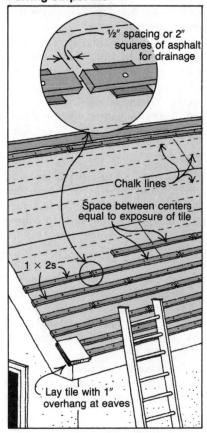

½" spacing or 2" squares of asphalt for drainage

Chalk lines

Space between centers equal to exposure of tile

1 × 2s

Lay tile with 1" overhang at eaves

mended exposure, simply mark the exposures along each rake, starting from the top edge of the furring strip near the eave. Snap horizontal chalk lines from rake to rake, and install furring strips with their upper edges along the chalk lines.

Where the roof doesn't divide into a whole multiple of the exposure, decrease the exposure slightly for all courses. (You should always decrease—never increase—exposures to compensate for the roof's irregularities.)

Marking chalk lines. If you're putting down tiles without furring strips, mark the roof deck with carefully placed chalk lines to guide you as you install each course.

First, lay a sample tile along the eave at each rake, allowing for the recommended overhang (1¼ to 3 inches). Snap a chalk line between the upper edges of the two tiles.

Next, measure the distance from the chalk line to the top of the

sheathing. Divide the distance by the exposure that the manufacturer recommends for your tile.

If the distance divides into a whole multiple of the exposure, then mark for successive exposures from your first chalk line to the ridge board, and snap horizontal chalk lines. Then, as you lay tiles, align their upper edges with the chalk lines.

If you have to compensate for an odd fraction of the exposure, do so by shortening tile exposures over the whole distance from eave to ridge.

Tips for cold climates

If January temperatures hover around or below 25°F/−4°C in your neighborhood, ice dams will form at your eaves, where repeated thawing and freezing occurs (see sketch below). And as ice dams build up, pools of water collect behind them. To eliminate the risk of water penetrating the deck, take these precautions when you prepare the roof deck:

Protecting asphalt roofs. To protect asphalt shingle roofs in an ice dam climate, first apply standard underlayment to the sheathing (see pages 52–53). Then cover the eave

Ice dam

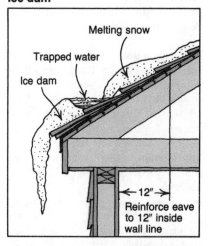

Melting snow

Trapped water

Ice dam

12"

Reinforce eave to 12" inside wall line

area—to 12 inches inside the exterior wall line of the house—with a 36-inch-wide sheet either of 90-pound mineral surface or minimum 50-pound smooth surface roll roofing. (See sketch next page.)

Allowing a ⅜-inch overhang along the eaves, nail down the roll roofing at intervals of 12 to 18 inches, 6 inches from the bottom edge and 1 inch from the upper edge. At vertical joints, allow 4 inches overlap; fasten the lower lap to the roof with nails every 12 inches, and cement the two ends together as illustrated.

Reinforcing the eaves

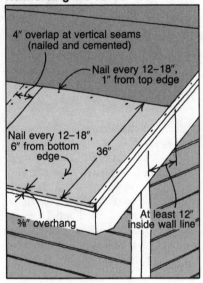

4" overlap at vertical seams (nailed and cemented)

Nail every 12–18", 1" from top edge

Nail every 12–18", 6" from bottom edge

36"

⅜" overhang

At least 12" inside wall line

If you need to apply more than one width of roll roofing to reach 12 inches inside the wall line, overlap horizontal courses at least 6 inches, using plastic cement liberally where they overlap. Finally, add a drip edge along the rake (see page 61).

Protecting wood roofs. For wood shingle and shake roofs, use either a double layer of 15-pound felt or a single layer of 30-pound felt along the eaves. Overlap the felt every 18 inches (overlap 4 inches at vertical joints), and cement the lapped layers together in the manner described for "Low-slope asphalt roofs," following.

Protecting tile roofs. Use a single layer of 40-pound or heavier coated felt, extending it from the eave to 24 inches inside the exterior wall line of the house. Then use 15-pound felt to the ridge.

De-icing tapes. Electrically heated cables, or de-icing tapes, also can be installed along the roof eaves to facilitate proper drainage of melting snow and ice.

The tapes, insulated for safety, are clipped to shingles in a zigzag pattern (see sketch below) and plugged into a waterproof electrical outlet. When heated (to about 86°F/30°C), they create miniature drainage channels for water that otherwise would back up behind an ice dam.

You can find de-icing tapes in 20 to 100-foot lengths at home improvement centers and roofing supply companies. Be sure to follow manufacturer's instructions for installation.

Snow guards. When it slides off the roof in glacierlike chunks, snow can tear gutters from their fastenings, rip away roofing materials, and half-bury the friend who just rang your front bell.

To prevent heavy, sudden snowslides like these, you can install metal snow guards in staggered rows

Cold climate hardware

De-icing tapes

Snow guard

over the roof. They can be nailed to the sheathing before new roofing materials are laid or hooked over nails underneath existing shingles.

How many snow guards you install depends on the slope of your roof. Generally, for every 100 square feet of roof, you should install 50 guards on a roof with a 6 in 12 slope, 75 guards on a roof with an 8 in 12 slope, and 125 guards on a roof with a 12 in 12 slope.

Tips for low-slope roofs

Plan to reinforce the underlayment if you are putting down a new surface over a roof with a slope under 4 in 12. Low-slope roofs shed water less easily than steeper roofs, and they present a weaker barrier against the horizontal movement of water that can back up underneath the shingles.

Low-slope asphalt roofs. Asphalt shingles can be applied to roofs with slopes down to 2 in 12 if two layers of 15-pound (or one layer of 30-pound) roofing felt are used in the underlayment. (If you have a low-slope roof, use only shingles with factory-applied mastic adhesive dots or lines on the upper edges. They seal one shingle to the next and prevent wind from lifting the tabs.)

When you apply the underlayment, cut a 19-inch-wide sheet of felt and nail it along the eaves, allowing ⅜-inch overhang along the drip edge (see sketch on next page). Overlap 4 inches at vertical joints.

Next, use a trowel or roofing brush to apply a continuous layer of plastic cement (at the rate of 2 gallons per 100 square feet) onto the felt.

Third, press a 36-inch-wide sheet of felt firmly onto the cement, and nail down the felt with 1¼-inch galvanized roofing nails placed every foot along a line 18 inches above the bottom of the felt.

Again, apply plastic cement to the upper 19 inches of the exposed felt. Cover the cemented portion with the next sheet of felt, and nail in the manner described above. Repeat the

Protecting low-slope roofs

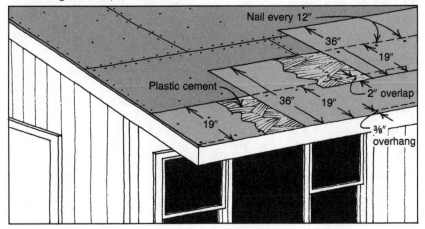

Nail every 12"

36"

19"

Plastic cement

36"   19"   2" overlap

19"

3/8" overhang

process until you are 24 inches inside the interior wall line of the house.

After that, put down double layers of the felt, without cement, up to the ridge.

Low-slope wood shake roofs. Your building inspector may approve the use of wood shakes for a roof with a 3 in 12 slope, though some professional roofers advise against it because the wood doesn't shed water easily and thus tends to deteriorate more rapidly, especially in hot or humid climates.

If you decide to install a wood roof over a low slope, add a layer of 15-pound felt underlayment under the shakes in addition to the 30-pound felt interlay.

Low-slope tile roofs. Your building inspector may also approve the installation of a tile roof over slopes as low as 3 in 12.

Underlayment for low-slope tile roofs is usually composed of a double layer of 15-pound felt, cemented together with plastic cement in the same manner as described above.

FLASHINGS

Usually made of rust-resistant galvanized sheet metal (see page 46), flashings protect the roof at its most vulnerable points: in the valleys, at the edges of roof vents, chimneys, and skylights, along the eaves of

the house . . . anywhere water can seep through broken joints into the sheathing.

If you're reroofing or working with a stripped deck, it may only be necessary to reseal flashing joints with plastic cement if you find the flashings still in good condition (see page 76). Should you need to install new flashings, the instructions here will guide you through the task.

Be sure to wear gloves when you're working with metal—especially rusted metal—to avoid cutting yourself on its sharp edges.

Valley flashings

Valleys in particular require sturdy flashing because they conduct more water to the gutters than do any of the roof planes.

Most valley flashings are made of galvanized metal, though mineral-surfaced roll roofing may be used for valleys on asphalt roofs. Finishes vary from "open" valleys, where shingles or tiles are cut away to expose the valley, to "woven" valleys, where asphalt shingles overlap to cover the valley.

When you're roofing over new or stripped decks, you can use 28-gauge galvanized metal for all sloping roofs, regardless of the roofing material you plan to apply. If you don't want to make your own, you can buy valleys from a roofing supplier to correspond to the slope of your roof. They often come with a 1-inch splash

guard at the center of the valley, and a crimp at the outer edges to direct water back to the center. The metal usually extends 8 to 11 inches up each side of the valley.

If you're roofing over asphalt shingles, you can also use 36-inch-wide roll roofing (color-matched to the new shingles) at the valleys.

Installing metal valleys

To install a flat-edged metal valley on either new or stripped decks or over old shingles, first lay a 36-inch-wide panel of 15-pound roofing felt along the valley's length, as the sketch below indicates. Then, using 1¼-inch galvanized nails, nail down the metal at 12-inch intervals along both edges, driving nails 1 inch from the outer edges. (Never nail in the center of the valley; holes there will allow water to penetrate sheathing.)

To install crimp-edged valleys, use cleats every 12 inches along the edges and fasten them with nails.

Installing metal valleys

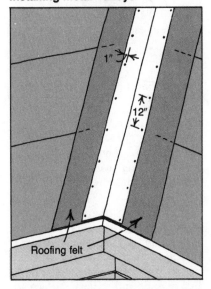

1"

12"

Roofing felt

Next, for both appearance and protection, spray paint the metal to match the color of your roofing materials. Use a rust-resistant metal paint available from roofing suppliers or home improvement centers. After cleaning the metal with solvent, apply a zinc-based metal primer. Then spray on two light coats (rather than one heavy coat) of paint.

Finally, finish the valley according to one of three methods described under "Valley finishes" following.

Installing roll roofing valleys

To install valleys made of mineral-surfaced roll roofing on either new or stripped decks or old roofs, first lay a 36-inch-wide panel of 15-pound roofing felt down the length of the valley. On top of that, lay a continuous sheet of 18-inch-wide roll roofing, mineral surface down. (If you

Installing roll roofing valleys

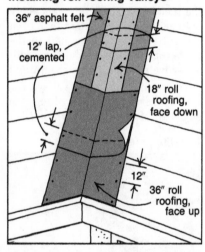

36" asphalt felt

12" lap, cemented

18" roll roofing, face down

12"

36" roll roofing, face up

must use more than one piece to cover the length of the valley, overlap 12 inches where pieces meet; see below.) Using galvanized roofing nails, nail along the outer edges every 12 inches.

Next, trowel liberal amounts of plastic roofing cement (2 gallons per 100 square feet) over the layer of roll roofing. Then lay another continuous strip of roll roofing, this time 36 inches wide and mineral surface up, over the cement.

Valley finishes. If you have an asphalt roof, you can complete valleys in one of three ways—with "woven," "closed-cut," or "open" finishes (see illustration above right). If the roof is wood or tile, "closed" and "open" finishes are suitable.

Three valleys

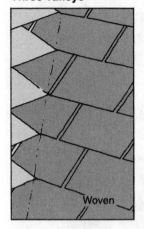 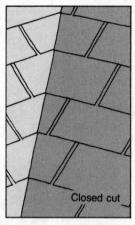 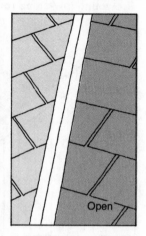

Woven

Closed cut

Open

Woven valleys, sometimes preferred for their appearance and the exceptionally good weather protection they offer, are made with asphalt shingles that overlap one another over the valley.

To make a woven valley, work from the eave to the ridge, extending courses of shingles beyond the center of the valley and up the adjoining slope, as the sketch below indicates.

When nailing the last shingle of each course, drive nails at least 6 inches from the center of the valley. If a shingle ends in the middle of the valley, cut off one tab and use another whole shingle to clear the distance. Press the final shingle tightly into the valley.

Next, extend the end shingle from the same course on the opposite slope over the valley. Weave the two

Weaving a valley

Nail at least 6" from valley center

end shingles together, one on top of the other. Alternate top and bottom shingles from opposite directions as you proceed with each course from eave to ridge.

Closed-cut finishes also camouflage valleys with asphalt shingles.

To make a closed-cut valley, extend all the shingles from one intersecting slope at least 12 inches beyond the center of the valley. Be careful to nail the end of each shingle at least 6 inches from the valley center (see illustration below).

Next, extend the shingles from the opposite slope to the center of the valley. Trim each shingle at least 2 inches from the valley center, following a line that parallels the center of the valley. Work from the eave to the ridge.

Secure the cut shingles to the cen-

A closed-cut valley

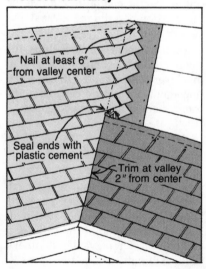

Nail at least 6" from valley center

Seal ends with plastic cement

Trim at valley 2" from center

ter of the valley with liberal amounts of plastic cement.

Open valley finishes are those in which the shingles (asphalt or wood) or tiles have simply been trimmed along the outer edges of the flashing.

To make an open valley with asphalt, wood, or tile, snap two chalk lines the full length of the valley, one on each side of the center (see illustration below).

Open valleys: snapping a chalk line

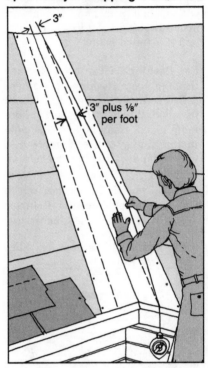

The chalk lines should start at the ridge 3 inches from the center on each side of the valley. As they continue toward the eave, however, both lines should slant away from the valley center ⅛ inch per foot, to allow for the additional volume of water and debris the valley will carry as it nears the gutter. If, for example, the valley is 8 feet long, the chalk lines should begin 3 inches from the valley center at the top, and end 4 inches from it at the eave.

Next, bring each course of shingles or tiles to the chalk line as shown above. Using a utility knife for asphalt shingles, a hand saw or roofer's hatchet for wood, or a circular saw with a carborundum blade

Finishing an open valley

for tiles, trim materials at the chalk line.

Closed valley finishes are achieved when wood or tile courses are trimmed to meet flush at the center of the valley. If you're finishing valleys this way, use a roofer's hatchet to trim wood and a carborundum blade on a circular saw to trim tile,

Chimney flashing

and drive the final securing nails as far as possible from the valley center.

Chimney flashing

A chimney is the most difficult part of the roof to make and keep waterproof, primarily because of its weight. The weight of a masonry chimney causes it to settle and move independently of the house.

Because water can so easily penetrate broken joints between a chimney and a roof, two layers of flashing are often installed—"base" flashing underneath and "cap" (or "counter") flashing on top (see page 60).

If you're roofing with asphalt, you can use either galvanized metal or roll roofing for chimney flashing. If you're roofing with wood or tile, use metal. Lead may also be used for tile roofs and is often preferred because it conforms easily to the shape of the tiles.

If you don't want to make your

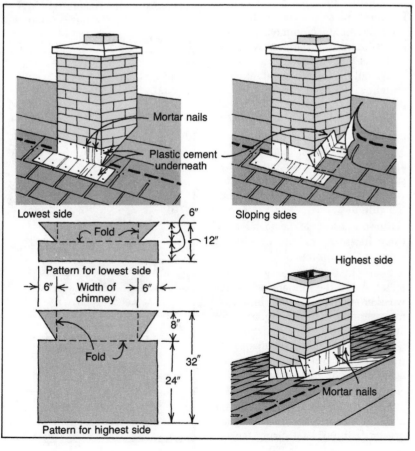

own chimney flashings, look for ready-made metal chimney flashings at roofing supply companies or sheet metal shops. A sheet metal shop may also be able to fabricate chimney flashing to the specifications you provide.

Base flashing for new and stripped decks. Whatever material you use to make chimney flashings, cut the patterns to fit around the chimney, as the sketch on page 59 indicates.

Before you begin, clean the chimney surface with a wire brush on all four sides, at least 6 inches above the deck on the lowest and sloping sides, and 12 inches high on the highest side. Apply an asphalt primer, available at roofing supply stores, to the cleaned areas. Then apply plastic cement liberally over the cleaned areas.

Lowest side: Cut a strip of material 12 inches wide and 12 inches longer than the width of the chimney at its lowest side. Complete the bends and cuts as illustrated on page 59.

To install the lowest side cutout, first apply plastic cement to the deck. Then, laying the cutout on top of the cement, nail it to the deck with 1¼-inch roofing nails. Using mortar nails, nail the top of the cutout to the chimney. (Proceed carefully so you don't crack the mortar.)

When bending the cutout to fit against the chimney, bend it to an angle slightly larger than the actual angle between the roof and chimney surfaces. The resulting spring tension will produce a tight fit against the chimney.

Sloping sides: Make metal steps (Step flashing) for the two sloping sides of the chimney.

First cut squares of galvanized metal twice the dimension of the weather exposure for the shingles—for example, 10-inch squares for shingles with 5-inch exposures. You'll need as many squares as the number of courses that run up against the chimney on both sloping sides.

Bend each square into the shape of an L, making the angle slightly greater than 90° so a spring tension

Step flashing

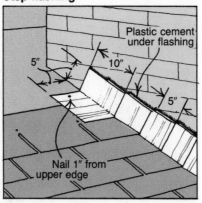

will hold the metal against the chimney.

Install the steps over each end shingle where courses abut the chimney, as the sketch above indicates. Use liberal amounts of plastic cement underneath each metal step and between the metal step and the chimney. Then fasten the flashing to the deck with a single 1½-inch galvanized roofing nail, driving the nail 1 inch from the upper edge of the step.

Highest side: Install the highest side last. Because more water will

Cricket flashing

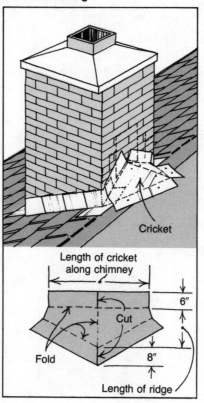

Cricket

Length of cricket along chimney

Cut

Fold

6"

8"

Length of ridge

flow against the chimney on the high side, make this flashing 32 inches wide and 12 inches longer than the width of the chimney. The 32-inch width will allow you to cover 8 inches of the chimney and 24 inches of the deck above the chimney.

Fold and cut the flashing as illustrated on page 59.

If you find a cricket or saddle at the back face of the chimney (it prevents snow and ice from accumulating, and it deflects water), cover it with asphalt primer and cement and install flashing according to the pattern illustrated below left.

Cap flashing. Over metal chimney flashing, cap flashing (sometimes called counter flashing) may also be installed. It consists of sheets of durable metal, often copper, installed in the mortar of the chimney and then bent down over the four cutouts.

To prepare the chimney for cap flashing, chisel the mortar from a horizontal joint above the flashings; chisel to a depth of 1½ inches (see illustration below).

Cap flashing

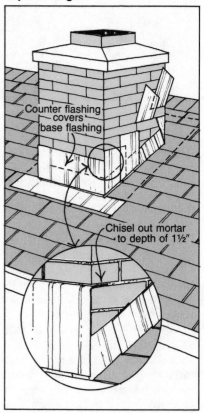

Counter flashing covers base flashing

Chisel out mortar to depth of 1½"

To install the cap flashing, it is easiest to use the "instant" mortar available at home improvement or hardware stores. Mix the mortar thoroughly, adding only enough water to form a thick paste. Wet the chiseled area well, and trowel the mortar into the chiseled area. Then push the cap flashing into the wet mortar and hold it there for several minutes while the mortar hardens. Wait a day for the mortar to set before bending the flashing into position.

Flashing for reroofing. When roofing over asphalt or wood shingles, leave all old shingles in place. Leave chimney flashings in place if you find them still in good condition. If not, make new flashings in the manner described, using the old flashings as patterns.

Skylight flashing

Skylight units are self-flashing, but the wood box on which a skylight rests (it's called a "curb") must be flashed (see pages 78–79).

The curb is usually 6 inches high. Flash the curb by modifying the instructions for the chimney: the height of the cutouts will be less than for a chimney, and with a curb, you needn't install counter flashing.

Vertical wall flashing

To flash where a sloped roof meets a vertical wall, you can install L-shaped metal that's crimped at the edges—if it will later be covered with siding. If not, you can take another approach with mineral-surfaced roll roofing or sheet metal.

To install solid strips of L-shaped metal, position the strips and secure them with 1½-inch galvanized roofing nails driven 1 inch from the outside edge at 1-foot intervals.

When you install L-shaped steps, follow instructions for installing metal steps in chimney flashing (preceding).

If you're applying roll roofing or

sheet metal, fold a 12-inch-wide strip into the intersection between the roof and the wall (see illustration below) from the eave to the ridge, troweling on liberal amounts of plastic cement underneath the strip. Nail the strip with 1½-inch galvanized roofing nails every 6 inches up both sides, 1 inch from the edges.

Vertical wall flashing

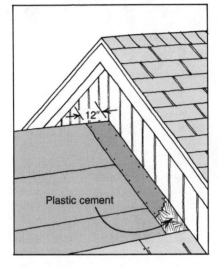

12"

Plastic cement

When the strip is installed, use a caulking gun to apply a thick bead of plastic cement at the seam where the strip meets the vertical wall. When you lay shingles that abut the strip, embed them in another coat of plastic cement laid over the strip.

Drip edges

Eaves and rakes require drip edges, which prevent water from seeping back up under the roofing material as it flows off the roof.

Along the eaves, install drip edges underneath the roofing felt. Along the rakes, install them on top. Using 1¼-inch galvanized roofing nails, fasten either preformed flashings or a 5-inch-wide strip of galvanized metal or aluminum along the length of each eave and rake, allowing a 2-inch overhang, as the sketch at right indicates. Nail every 10 inches, placing nails 1½ inches below the upper edge. Then bend the 2-inch extension downward.

Vent pipe flashing

Flashing around ventilation pipes is usually installed with the roofing materials, for part of the flashing fits over the shingles. The upper part is then covered by shingles (see illustration on page 62).

You can often buy metal or plastic flashings from roofing supply companies to fit protruding vent pipes for plumbing, appliances, and ventilation fans. Or you can make your own from sheet metal or asphalt roll roofing. Lead is commonly used for vent flashings on S-shaped tile roofs because it is pliable and will fit the curves of the tile.

Whether purchased or handmade, vent pipe flashings should have a base large enough to carry water away from the pipe to layers of roofing below. They should also have secure, waterproof seals where flashings and vent pipes meet.

Measure the diameter of the vents before buying or making flashing. The base of a vent flashing should be a square whose sides are 12 inches larger than the diameter of the vent.

To make your own vent pipe flashing, cut a piece of metal or roll roofing after the pattern illustrated on page 62.

Next, mark a line 4 inches above the bottom edge. Line up the mark with the down-roof side of the pipe, and make another mark even with the up-roof side of the pipe.

Then lay the flashing on the roof below the pipe, lining up the center

Installing drip edges

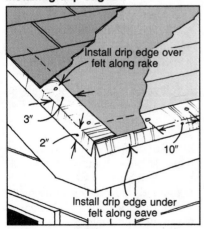

Install drip edge over felt along rake

3"

2"

10"

Install drip edge under felt along eave

Vent pipe flashing

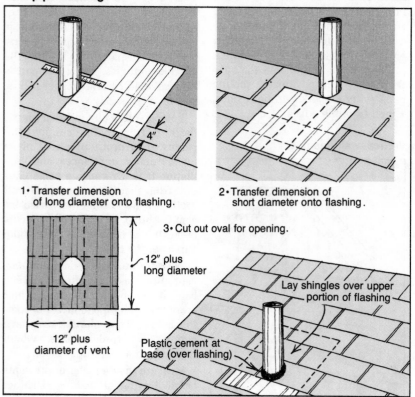

1• Transfer dimension of long diameter onto flashing.

2• Transfer dimension of short diameter onto flashing.

3• Cut out oval for opening.

12" plus long diameter

12" plus diameter of vent

Lay shingles over upper portion of flashing

Plastic cement at base (over flashing)

of the flashing with the center of the pipe. Mark two more lines even with the sides of the pipe. The rectangle in the middle will indicate where to cut an opening.

When you have the vent opening cut out, trowel on liberal amounts of plastic cement to the base of the vent and install the flashing over it. Apply more cement around the base of the vent pipe at the joint.

Cut additional roofing materials as needed to fit above and on either side of the flashing.

ROOF INSULATION

If you've been thinking of insulating your attic, consider tying the project into your roofing plans. Adding insulation to an unfinished attic or a beamed ceiling is relatively easy work, and it can cut your fuel bills by as much as 30 percent. This is particularly true of a one-story house, where most of the house's heat escapes through the roof.

How much insulation do you need?

How much insulation you should add depends both on the amount of insulation you already have in your house and on the R-value recommended for your climate. R-values indicate the degree of resistance insulation materials have to heat flow; the higher the R-value, the better its resistance. Check with your utility company to ascertain the recommended R-value for attic or ceiling insulation in your area.

Types of insulation

Rigid insulation boards or batts of fiberglass or rock wool are most commonly used in attic or ceiling insulation.

Rigid insulation boards. Lightweight rigid insulation boards—made of compressed fiberglass, polystyrene, or urethane—come in thicknesses ranging from ¾ inch to 4 inches. They're also available in 4

by 8, 4 by 4, and 2 by 8-foot panels.

In homes with unfinished attics, the panels are usually installed between attic floor joists. In homes with exposed-beam ceilings, they can be installed either between the ceiling rafters or on top of the roof deck before shingles are laid (see sketch below).

Rooftop insulation

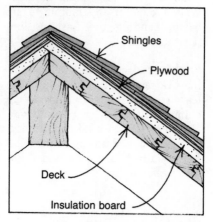

Shingles

Plywood

Deck

Insulation board

If you install insulation board on top of the roof deck, consider covering it with another layer of wood decking in order to provide a more stable nailing base for the shingles. Roofing material manufacturers generally prefer that insulation be installed underneath the roof deck. Asphalt shingles laid directly over insulation can't easily dissipate the heat of the summer sun; and wood shingles or shakes also last longer if not exposed, when wet, to the heat stored by insulation directly beneath them.

Fiberglass and rock wool batts. Blanketlike batts of fiberglass or rock wool are often fitted between floor joists or laced between the rafters in an unfinished attic.

Fiberglass batts are made from glass fibers; rock wool is spun from molten slag rock. Both are available in several thicknesses in 15 or 23-inch widths.

Installing insulation

You'll find specific instructions for applying insulation to your attic or ceiling in the *Sunset* books *In-*

sulation & Weatherstripping, and *Energy-saving Projects.*

If you plan to put down insulation over the roof deck, check the warranty on the roofing materials you plan to install; follow the manufacturer's recommendations.

ROOF VENTILATION

Showers, cooking, washing, and just breathing can cast a surprising amount of water into the air of your home—5 to 10 pounds a day. Add another 30 pounds of moisture if you wash and dry clothes.

Without proper ventilation through roof vents and attic fans, moisture condenses in the attic and eventually damages sheathing, rafters, insulation—and even asphalt or wood roofing materials. Proper ventilation, on the other hand, allows a house to "breathe" unwanted moisture, as well as accumulated heat, smoke, fumes, and vapors.

To improve the ventilation of your own house, make changes when you prepare the roof deck.

How much ventilation do you need?

The amount of ventilation your house needs depends on factors such as wind direction, sun and shade, and roof lines that may interrupt or encourage air flow.

Your building inspector or a ventilation contractor can help you determine whether your house needs additional vents and—if it does—their sizes and locations. (You can find contractors in the Yellow Pages under "Contractors, Heating and Ventilating.")

As a general rule, provide 1 square foot of free vent opening (an opening without wire or grillwork) for each 150 square feet of attic floor area. Subtract the area taken up by wire or grillwork. A vent covered by ⅛-inch or ¼-inch wire mesh should be 1¼ times as large as a free vent opening for the same area. A vent

covered by ¹⁄₁₆-inch insect screening (or ¼-inch mesh and a louver) should be twice as large.

A ratio of 1 square foot of free vent opening for each 300 square feet of attic floor space may be adequate if a vapor barrier is installed on the room side of the insulation and if half of the vent space is located near the tops of the gables or along the ridge.

Where natural venting may not provide adequate ventilation, an attic fan may be necessary to push hot air through vents.

Types of roof ventilation

On a roof, attic vents or fans are commonly located in the uppermost corner of a gable, underneath the eaves in soffits, on the roof plane, or on the ridge, as the illustration below shows.

Gable ventilators. Triangular gable ventilators, made of galvanized metal, are available at home improvement centers or sheet metal shops. Installed at the top of the gable, they eliminate heat that collects near the house ridge.

Soffit ventilators. Rectangular ventilators placed at the soffit or eave

Four ways to ventilate

area of the roof provide an in-flow of cool air. Warm-air convection then draws this air up to and through gable or ridge ventilators. Soffit ventilators help dry out roof decks that have leaked, particularly at the eaves.

Roof plane turbines and fans. Turbine vents placed on the roof plane act as free ventilating space in calm weather, but generate an air flow from the turbine action when the wind blows.

Powered attic exhaust fans augment natural air convection. Placed over ceiling vents, these fans can substantially reduce air conditioning needs in a hot climate.

When purchasing an exhaust fan, consider its air flow in terms of cubic feet per minute (CFM), and its noise level. Fans are rated for loudness in "sones," an international standard. (The lower the number of sones, the quieter the fan.)

Ridge ventilators. Because the hottest air collects at the ridge of the roof, the ridge is the most efficient single area for placing a vent.

Ridge vent systems consist of a long, inverted metal trough that allows free air flow out of the house without admitting rain. Ridge vents must be installed before roofing materials are laid.

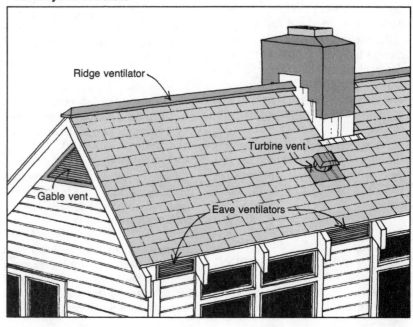

Ridge ventilator

Turbine vent

Gable vent

Eave ventilators

LAYING THE ROOF

With the groundwork—or rather, roofwork—prepared down to the flashings (see "How to Prepare the Roof Deck," pages 50–63), you're ready to lay the roofing. You'll find it's the easiest part of your project *if* you've taken care to measure and prepare the surface, planning your alignments to compensate for any irregularities.

Whether you're roofing with asphalt shingles, roll roofing, wood, or tile, the procedure boils down to a three-stage operation: first you lay a "starter" course to bolster the first row of materials; second, you apply materials, row by row, from the eaves to the ridge; and third, you cap the roof's ridges and hips with specially made shingles or tiles (or, if you're installing it, bands of roll roofing).

For tools, you'll need a hammer or roofer's hatchet, chalk line, tin snips (for asphalt shingles and roll roofing), a putty knife, a cutting knife or saw (for wood or tile), and, if you have one, a nail stripper loaded with roofing nails.

For materials, you'll need roofing shingles or tiles loaded onto the roof (see "When you order," page 48), as well as nails and a gallon of plastic cement.

SEVEN TIPS FOR ROOFING

Before you begin, familiarize yourself with the following procedures for proper nailing, working in a pattern, aligning shingles, and compensating near the ridge. Now is also a good time to refresh yourself on roofing safety (see pages 40–41).

Listen to your manufacturer

Read carefully—and follow—the instructions that will come with whatever roofing materials you buy. Manufacturers' directions may vary slightly from those provided in the following pages. To ensure that your warranty will be in force, stick to manufacturers' recommendations.

Where to begin

When you start laying the roofing, begin on the side that is most prominent visually—here is where you'll have the most uniform appearance. After that, whether you begin at the right or left corner (or in the center of the roof) depends on the type of roof you have, on the materials you're using, and on your right-handed or left-handed orientation.

Gable roofs. If you have a gable roof, begin at the left rake if you're right-handed—starting here allows you to swing your right arm in an easy circle as you work. If you're left-handed, start at the right rake.

If you're putting down interlocking tiles, you may have to start at the right rake in order for the tiles to hook together properly.

Hip roofs. Since hip roofs have no rake, begin at the center by snapping a vertical chalk line equidistant from the corners. Then work from the center in both directions.

Starter course safety

You can apply the starter and first courses of shingles while you're on the roof, but be careful of becoming dizzy—and falling off the roof—while looking down at an angle.

You might feel safer if you work from a ladder or scaffold. Remember to keep your hips between the ladder rails as you work.

How to work in a "fan"

For economy of movement, once the starter and first courses are down, don't shingle one layer all the way across the roof. Stay at the starting point and lay a fan-shaped spread of shingles (the first five or so courses). Then move sideways on the roof and start another fan. (Working this way with asphalt shingles also reduces the problem of slight color variations from bundle to bundle.)

How to nail

Proper fastening is an essential part of good roofing. It depends on three things: selection of the right type and length of nail (see "Nails," page 46), use of the proper number of nails, and correct positioning (for asphalt shingles, see next page; for wood shingles, see page 69).

Aligning shingles and tiles

If you're roofing over older materials, new shingles may conveniently line up with the butts of the old ones. On new roofs, though, it's usually necessary to snap chalk lines to keep the shingles or tiles properly aligned.

Proper placement of chalk lines differs with each material; tiles require the most precise marks.

Compensating near the ridge

Consider yourself fortunate if all the courses meet evenly at the ridge. Chances are they will not.

Before you lay materials, measure your roof to see if it is the perfect rectangle it appears to be. If not, compensate gradually either by shortening the exposure on courses as you near the ridge or by trimming materials at the least obvious rake.

APPLYING ASPHALT SHINGLES

If you're roofing over old shingles, you can usually line up new shingles with the butts of the old ones. If roofing over a new deck, line up shingles with vertical and horizontal chalk lines. If you're putting down metric-size shingles, follow the manufacturer's instructions for both exposure and alignment; metric shingles are a little larger than standard shingles, so your measurements will differ slightly.

Before you begin to lay the shingles, see that the roof deck has been properly prepared with underlayment and, if necessary, new flashings (see pages 50–62).

Exposure

The correct weather exposure for most asphalt shingles is 5 inches. That means that 5 inches of the bottom of each shingle will be exposed to the weather after overlapping courses are applied.

Nails and nailing

For asphalt shingles use 12-gauge, galvanized nails with ⅜-inch-diam-

eter heads. Some roofers use nonferrous aluminum or stainless steel.

For new asphalt roofs, use 1¼-inch nails. Use 1½-inch nails when roofing over an old asphalt roof, and 1¾-inch nails when roofing over an old wood roof.

Nails are usually positioned 1 and 12 inches from the ends of each shingle, 3 inches from the bottom edge at the starter course, and 5⅝ inches above the butt line on the rest of the courses. (You'll find that most asphalt shingles are premarked with cutouts for accurate nailing.)

How to nail asphalt shingles

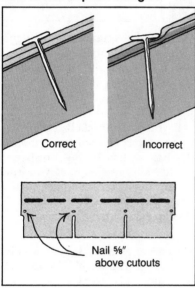

Correct   Incorrect

Nail ⅝"
above cutouts

Starter course, first course

The narrow starter row of shingles runs the length of the eave to make a base for the first full course of shingles. It also provides a solid surface for the shingle cutouts that will lie on top.

Before you lay the starter and first courses of shingles, you should have installed drip edges along the eaves and rakes (see page 61).

Laying the starter course. Measure the length of the eaves. Then select enough 36-inch-long shingles to cover the distance.

If you're reroofing, make the starter course 5 inches wide to correspond to the exposure of the lowest course

of shingles on the old roof (see illustration below). Cut 5 inches off the tabs and 2 inches from the top edges of 12-inch-wide shingles. (To cut shingles, use a utility knife and a straightedge or carpenter's square. Score the back of the shingle, then bend it to break.)

For a new roof, apply a 9-inch-wide starter course. Either use a 9-inch-wide strip of roll roofing, or cut 3 inches off the tabs of 12-inch-wide shingles.

Starter course: asphalt shingles

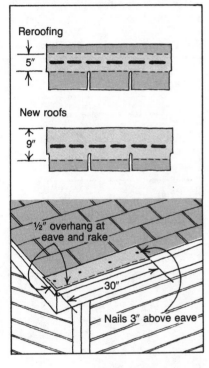

Reroofing

5"

New roofs

9"

½" overhang at eave and rake

30"

Nails 3" above eave

Starting at the left rake, apply the starter course along the eave with the wind sealing strips down. Trim 6 inches off the end of the first shingle to offset the cutouts in the starter and first courses.

Allowing a ½-inch overhang at both the eave and the rake, and 1/16-inch spacing between shingles, fasten shingles to the deck using 4 nails placed 3 inches above the eave. Position the nails 1 and 12 inches from each end.

Nail the starter course across the eave.

Laying the first course. When reroofing, you'll need a 10-inch-wide first course to cover the two 5-inch exposures of the first two courses of old shingles. Cut 2 inches off the top edges of as many shingles as were necessary for the starter course.

For new roofs, use full-width shingles for the first course.

Allowing the same ½-inch overhang at the rake and eave, and 1/16 inch between shingles, nail the first course over the starter course using 4 nails per shingle. Space the nails 5⅝ inches above the butt line, and 1 and 12 inches from each end, or according to manufacturer's instructions.

First course: asphalt shingles

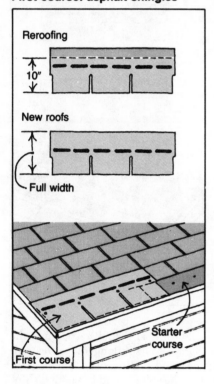

Successive courses

Your main concern when you lay the second and successive courses is proper alignment of the shingles—both horizontally and vertically. Horizontally, you'll lay shingles either against the butts of old shingles or along newly marked chalk lines. Vertically, you'll arrange shingles so they produce the pattern—centered, diagonal, or random—you want your roof to have.

Chalk lines for asphalt shingles

Horizontal alignment. Aligning shingles horizontally is simply a matter of placing new shingles against the butts of old ones (reroofing) or snapping chalk lines across the deck (new roofs).

If you're using chalk, snap lines every 10 inches from the bottom of the first course as the illustration above indicates. Then, as you work toward the ridge, the upper edge of every other course of shingles should line up against the chalk marks.

Vertical alignment, three-tab shingles. If you're working with standard three-tab shingles, you can produce one of several roofing patterns through different vertical arrangements of the shingles. To produce centered, diagonal, or random roof patterns, all you do is adjust the length of the shingle that begins each course.

Before you start your second row of shingles, snap vertical chalk lines from the roof ridge to one end of every third shingle along the first course. (See sketch above.) For extra accuracy in centered alignment, you may wish to snap a chalk line at the ends of each shingle along the first course.

Then start laying courses according to one of the following patterns.

Centered alignment offers the most uniform roof appearance, but it's also the most difficult pattern to work with because cutouts or shingle edges must line up—within ¼ inch—with cutouts or shingle edges two courses above and below (see illustration at right).

If you want to center the shingles, cut 6 inches or half a tab off the shingle at the rake in the second course, 12 inches or one tab in the third course, 18 inches or one and a half tabs in the fourth course, 24 inches or two tabs in the fifth course, and 30 inches or two and a half tabs in the sixth course. Work six courses at a time, setting the shingles at the rake, then working in a fan-shaped pattern. Allow 1/16 inch between shingles. After the sixth course, repeat the sequence, starting again with a full shingle.

Diagonal alignment is a little more forgiving if slight errors are made in calculating.

To align shingles diagonally, cut 4 inches off the second, then 8 inches off the third course, after the pattern illustrated at right. Then start again in the fourth course with a full shingle. Repeat the pattern to the ridge.

You can also produce a diagonal pattern with cutoffs in multiples of 5 inches in courses two through six.

Random alignment is the easiest of the three patterns to achieve, and it produces a more rustic appearance in the roof surface.

To lay shingles in a random pattern, trim the left tabs of the rake shingles for courses two through five to widths of 6 inches, 9 inches, 3

Three patterns for asphalt shingles

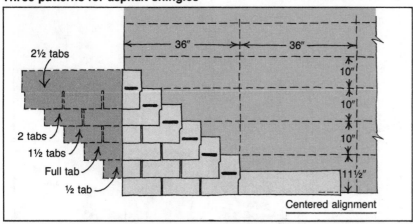

2½ tabs

2 tabs

1½ tabs

Full tab

½ tab

36″   36″

10″

10″

10″

11½″

Centered alignment

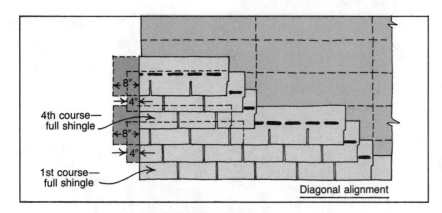

4th course—
full shingle

8″

4″

8″

4″

1st course—
full shingle

Diagonal alignment

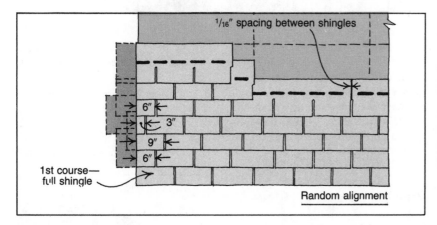

¹⁄₁₆″ spacing between shingles

6″

3″

9″

6″

1st course—
full shingle

Random alignment

inches, and 6 inches respectively (see sketch above). Then start again with a full shingle and repeat the pattern. Continue to the ridge, allowing ¹⁄₁₆-inch spacing between shingles.

Vertical alignment, premium shingles. Though premium shingles are often designed for a random, three-dimensional appearance, recommended methods of vertical align-

ment vary from one manufacturer to the next. If you're putting down premium shingles, follow instructions that come with the materials.

Applying hip and ridge shingles

If you haven't purchased ready-made ridge and hip shingles, cut 12-inch

squares for the roof's hips and ridges from 12 by 36-inch shingles (see "How to estimate ridge and hip shingles," page 47). Bend each shingle to conform to the roof ridge. (In cold weather, store shingles in a heated room to make them pliable.)

Also cut enough 7 by 12-inch pieces to use under the first shingles you nail to the ridge and hips.

Then, before you begin, snap a chalk line the length of the ridge and each hip, 6 inches from the edge.

Shingling the hips. If your roof has hips, start with them; put down ridge shingles afterward.

Starting with a double layer of shingles at the bottom of the hip, work toward the ridge, applying shingles with a 5-inch exposure. The edge of each shingle should line up with your chalk mark.

Use two nails, one on each side, 5½ inches from the butt and 1 inch from the outside edge (see sketch).

Shingling hips and ridges

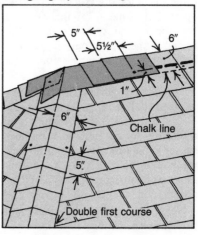

5″

5½″

6″

1″

6″

Chalk line

5″

Double first course

To finish the hips, cut 4 inches up the middle of a shingle. Fold these ends as illustrated, overlapping them when you fit the shingle to the roof. Nail at the overlap.

Shingling the ridge. Start at the end opposite the prevailing wind. Using nails that are long enough to penetrate the ridge beam securely (approximately 2 inches), apply the shingles in the manner just described. Dab the exposed nailheads of the last shingle with cement.

APPLYING ASPHALT ROLL ROOFING

Rolls of 36-inch-wide asphalt roofing are applied directly over the sheathing, with the 17-inch-wide strip of mineral surface exposed and the 19-inch-wide selvage, or uncoated portion, lapped underneath. Each layer is bonded to the next with plastic roofing cement.

Starter strip, first strip

Start with a 19-inch-wide strip (long enough to cover the distance from rake to rake) that has been cut from the roll (see illustration at right). If necessary, overlap vertical joints 6 inches. Fasten one lap to the roof with nails spaced every 4 inches (1 inch from the edge); then cement the other lap to the first.

Lay the starter strip along the eave with ½-inch overlap at the eave and rake, and fasten it to the deck with three rows of nails spaced 12 inches apart. Drive nails 4¾ inches from the upper edge, 4 inches above the bottom edge, and along the middle between them. Stagger nails as shown.

For new roofs, use 12-gauge galvanized nails with ⅜-inch-diameter heads and 1-inch-long shanks. For reroofing, longer nails may be necessary to penetrate old materials and reach either ¾ inch into a wood deck or completely through a plywood deck.

Next, following the manufacturer's specifications for rate of application (usually 1½ gallons per square), spread plastic roofing cement of a brushable consistency over the starter sheet. (Be careful not to apply too much, or it will ooze onto the surface when the next layer is applied.) Use an inexpensive roofing brush you can later toss out.

Then overlay a 36-inch-wide sheet and nail along the top (uncoated) 19-inch portion; place the nails in two rows, the first 4¾ inches below the upper edge and the second 8½ inches below the first row (see sketch at right).

Starter course: roll roofing

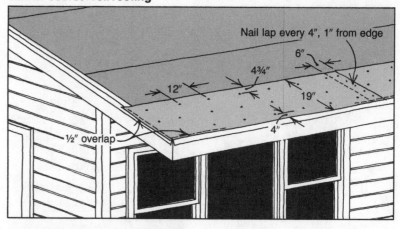

Successive courses: roll roofing

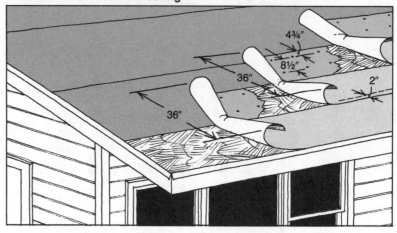

Successive strips

Bond each layer of roll roofing together with cement in the manner just described, continuing toward the ridge. Take care to seal vertical seams and to apply proper amounts of cement with each course.

Finishing hips and ridges

Cut enough 12 by 36-inch rectangles from rolls to cover hips and ridges (see "How to estimate hip and ridge shingles," page 47). Also cut enough 12 by 17-inch rectangles so you can double the starter shingles at each hip and ridge. Then snap a chalk line the length of the hip or ridge 5½ inches from the edge (see sketch at right).

Bend pieces lengthwise and, starting at the bottom of the hips or the end of the ridge opposite the prevailing wind, fasten them as shown below.

The upper, 19-inch portion of each shingle is fastened to the roof with nails placed every 4 inches, 1 inch from the outside edge. The lower, 17-inch portion is cemented in place.

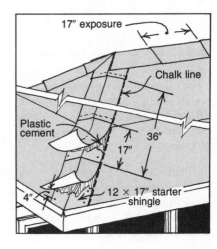

APPLYING WOOD SHINGLES AND SHAKES

When applying either wood shingles or shakes, always position the tapered end uproof and the thicker end downroof. If the wood product has a sawn side and a rough side, apply with the rough side exposed to the weather. When applying straight-split shakes (those equally thick throughout), apply with the smooth end uproof.

Before you apply shingles or shakes, prepare the roof deck and install flashings according to instructions on pages 50–62.

Exposure

Correct exposure for wood shingles and shakes depends on the length of the material and the slope of your roof. Here are recommended exposures:

	Exposure	
	3 in 12 to 4 in 12 slopes	4 in 12 and steeper slopes
Shingles		
16″	3¾″	5″
18″	4¼″	5½″
24″	5¾″	7½″
Shakes		
18″	—	7½″
24″	—	10″

Nails and nailing

For wood shingles and shakes, use corrosion-resistant nails, two per shingle or shake.

Nails for wood shingles should be 14½-gauge with 7/32-inch-wide heads. Use 1¼-inch nails for a new roof of 16 or 18-inch shingles, and 1½-inch nails for a new roof of 24-inch shingles. Longer nails may be needed to penetrate old roof surfaces and reach ¾ inch into or through the deck.

Locate the nails ¾ inch from the side of the shingle, and 1 inch above the butt line for the next course (see sketch above right).

For wood shakes, use 13-gauge nails with 7/32-inch-wide heads. Use 2-inch nails, unless longer nails are required to penetrate ¾ inch into the wood board deck or all the way through plywood.

How to nail shingles or shakes

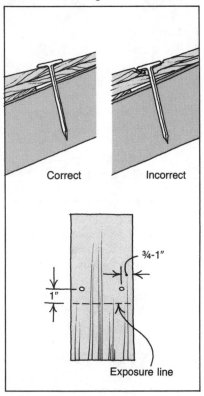

How to trim shingles and shakes

Occasionally you'll need to trim shingles or shakes as you reach a ridge or valley.

Making straight cuts. Cut shingles straight with the grain simply by slicing through them with a roofer's hatchet. Heavier wood shakes can either be sawn or split (with the grain) with a hatchet.

Making angled cuts. When it's necessary to make an angled cut for a valley, lay the shingle in place and use a straightedge to mark the desired angle of the cut. Then score the shingle with the hatchet and break it against a hard edge. Wood shakes must be sawn after they have been marked.

Starter course, first course

Combine the starter and first courses by laying the shingles or shakes one on top of the other (see illustration below). Overhang this double course 1½ inches at the eaves and rakes.

If you're roofing with wood shakes, though, first nail a 36-inch-wide strip of 30-pound felt along the eave (allow ⅜-inch overhang).

When laying the double course, offset the joints of the bottom layer by 1½ inches when you put down the upper layer. Allow ¼-inch spacing between wood shingles and ½-inch spacing between wood shakes to permit the wood to expand and contract without buckling.

Nail ¾ inch from the sides of wood shingles, 1 inch from the sides of shakes, and 1 inch above the butt line of the next course for both.

Applying wood shakes

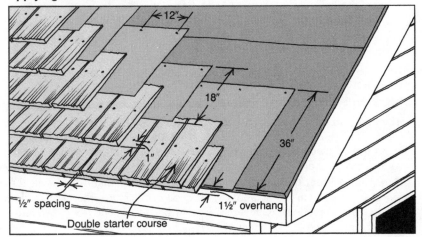

Successive courses

When you lay successive courses, align the shingles or shakes both vertically and horizontally for proper shingle exposure and coverage.

Vertical alignment. You don't need to snap chalk marks when you line up the shingles or shakes vertically. Simply lay the wood materials of random widths according to this principle: offset joints by at least 1½ inches so that no joints in any three successive courses are in alignment (see sketch below).

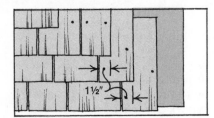

Horizontal alignment for wood shingles. To line up wood shingles horizontally, snap a chalk line at the proper exposure over the doubled starter/first course, or use your roofer's hatchet as an exposure guide (see sketch below). Then lay the butts

Horizontal alignment: shingles

of the next course of shingles at the chalk line, as the illustration above right indicates.

Nail the course, and repeat the process until you reach the ridge.

Horizontal alignment for wood shakes. Install felt interlays over each course of wood shakes as you work toward the ridge.

Here's how: From the butt of the starter/first course, measure a distance twice the planned exposure. Place the bottom edge of a strip of

Applying wood shingles

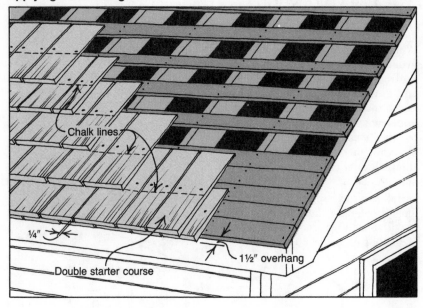

Chalk lines

¼"

Double starter course

1½" overhang

18-inch-wide 30-pound felt at that line and nail every 12 inches along the top edge of the felt. Overlap vertical joints 4 inches.

Then snap a chalk line on the starter/first course for the proper ex-

Horizontal alignment: shakes

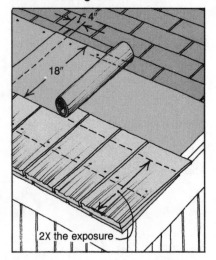

4"

18"

2X the exposure

posure (mark and use your hatchet handle as a guide).

Nail the second course, place the next felt, and continue until you reach the ridge.

Use short, 15-inch shakes as the last course. They either are ready-made or can be cut from standard-length shakes.

Applying hip and ridge shingles

Using factory-made ridge and hip shingles (beveling and making your own specialty shingles is a time-consuming business), double the starter courses at the bottom of each hip and at the end of the ridge as indicated in the sketch below.

Exposure should equal the weather exposure of the wood shingles or shakes on the roof planes. Start the ridge shingles at the end of the ridge opposite prevailing winds.

Use nails long enough to penetrate the layers of material and into the ridge board (usually 2 or 2½ inches).

Hip and ridge shingles and shakes

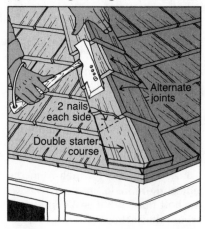

Alternate joints

2 nails each side

Double starter course

APPLYING TILE ROOFS

Because tile roofing materials haven't the same standard measurements as wood or asphalt materials (see "Clay and concrete tiles," page 45), they must be installed strictly according to manufacturers' instructions. Also, because tiles are heavy and difficult to handle, you'll find that most tile roofs are installed by professionals.

Most important to the job is proper layout of either chalk lines or furring strips (see pages 54–55). They must be spaced according to the recommended exposure for the tile.

Before laying tile, be sure that the roof is clean and that the interlocking undersides of all the tiles are free of foreign matter. Structural framing must be completed, if it's needed to support the weight of materials.

When walking on tile, prevent cracking by placing your foot parallel to the eave on the lower few inches of the tile exposure. If the tile is curved, walk on the ridges rather than in the troughs.

Exposure

Most 17-inch-long tiles have a 3½-inch overlap, allowing a 13½-inch exposure. Tiles overlap 1 inch or more along their side edges.

Nailing

Most clay tiles are either nailed or wire-fastened directly to the roof. Concrete tiles are either nailed to the roof or hooked over furring strips. In windy regions tiles must usually be nailed (check with your building inspector).

Tile nails should be 11-gauge corrosion-resistant box nails, 3 inches long or sufficiently long to penetrate ¾-inch into a wood board deck or all the way through a plywood deck.

How to cut tile

Cutting or trimming tiles—for valleys, vent pipes, hips, ridges, and rakes—requires a carborundum blade on a circular saw or a special water-cooled diamond blade. Be sure to wear safety glasses when you work.

Mark each tile with a guideline so you'll know where to cut the pieces to fit valleys and vent pipes. Then place the tile on a plywood board and cut the tile along the guideline.

'Birdstop' flashing

If you're putting down curved tiles, install "birdstop," or eave closure, flashing at the eaves before you lay your starter course. Designed to fit the curves of the tile (see illustration below), this flashing prevents birds and insects, as well as moisture and dirt, from entering the roof under the eave. (It may be required if you live in a fire-hazardous area.) It also supports the bottom row of tiles.

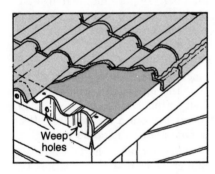

The type of birdstop flashing you install—usually individual ceramic pieces or 10- to 12-foot-long strips of metal—depends on what each manufacturer provides with the tiles you're using.

Laying tile courses

Whether you install tiles over furring strips or along chalk lines, you'll find that half the job is done when you have prepared the deck (see pages 54–55).

Interlocking devices on the sides of many tiles require that you begin at the right rake, though how courses are actually laid will depend on the manufacturer's recommendations and the type of tile you're installing.

Rake tiles

When curved or L-shaped rake tiles are used, they're installed either as each course is laid (see sketch below) or after all roof tiles are in place.

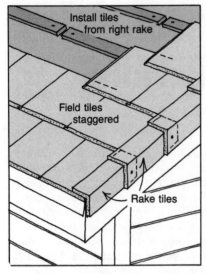

These tiles are designed both to protect the sheathing and to give the roof a finished appearance.

Hip and ridge tiles

Each tile manufacturer also produces specially-shaped hip and ridge tiles. As the illustration below indicates, they are usually mortared into place or fastened with nails and plastic cement over hips and ridges that have been built up with ridge or hip nailers (see page 54).

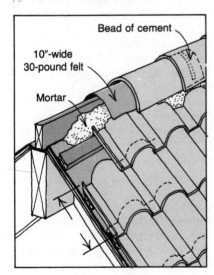

HOW TO MAKE ROOF REPAIRS

Considering the afflictions suffered by the ordinary roof — exposure to the extremes of summer and winter temperatures, pelting rains, hail, snow and ice, high winds, falling tree branches, kids in pursuit of runaway frisbees—it's surprising that roofs don't fail more often than they do.

At one time or another, unfortunately, most roofs do give way to the weather. One storm too many will create a leak along flashings, at the eaves, or where shingles have worn thin or tiles have cracked. A stiff wind will uproot loosened asphalt or wood shingles and scatter them over the lawn. The weight of accumulated snow and ice can pull entire gutter systems away from the roof, and the water that collects behind an ice dam can eventually seep into and erode sheathing.

To protect against major roof damage, it's best to correct minor failures as soon as you discover them. Inspect your roof periodically (see pages 35–38). Are the seals along the flashings in good order? If not, apply new sealants. Are there patches of worn shingles or broken tiles? Replace them. Are the gutters and downspouts free from debris? If not, take an afternoon to clean them out.

In the following pages you'll find instructions for making such minor roofing repairs, as well as instructions for finding and locating leaks, and correcting damage done to sheathing and rafters when a leak has been neglected too long.

FINDING & FIXING A LEAK

High on the list of irritations that we can all live without is the sound of water plunking rhythmically from the ceiling into pots and pans spread about the floor.

Unfortunately, your roof usually springs a leak when you're least able to deal with it—during a rainstorm. In the first place, crawling around on a wet, sloping roof is dangerous business. On top of that, a leak can't really be properly repaired until the roof is partially dry.

There are, nonetheless, certain emergency measures you can take to plug the source of a leak while you wait for the weather to change. When things dry out a bit, you can effect a more permanent solution.

To make even a short-term repair, though, you first have to find the leak—at its source.

Finding a leak

Leaks in a roof are mischievous: they have an uncanny knack for showing up far from the point where they originated. It's entirely possible, for instance, for water to enter a house through a crack along the living room chimney and slide sideways or diagonally between layers of roofing materials and down rafters until, perhaps at a nail tip or knot hole, it drops and collects in a puddle on a bedroom floor.

Most leaks begin at a roof's most vulnerable spots—at seals along flashings; on the roof plane where shingles or tiles are missing, torn, or cracked; in valleys that may be clogged by debris; or at the eaves, where standing water can penetrate the sheathing.

To locate and repair a leak, first examine the underside of the roof from the attic or crawl space, and make temporary repairs from there. Later, when the roof has dried out some, look over the roof surface and make more permanent repairs.

Examining the roof's underside. Working from the attic or crawl space, use a strong flashlight, a thin screwdriver, and a scraping knife to examine vents, skylights, vent stacks,

Locating a leak in the roof

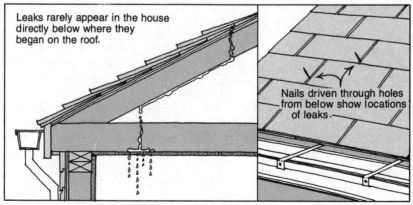

Leaks rarely appear in the house directly below where they began on the roof.

Nails driven through holes from below show locations of leaks.

chimneys, and valleys. Look for water spots, moisture, and soft spots that may indicate dry rot.

Also examine the ridge beam, rafters, and sheathing for dark-colored areas that signal wet wood. Mark wet spots with chalk so you can find them easily later on.

If it is necessary to remove fiberglass insulation batts to examine the sheathing, be sure to wear loose clothing, gloves, goggles, and a respirator for protection. (Damp batts of fiberglass, if laid in the sun to dry and put back in place, will still provide effective insulation. Loose-fill insulation, if dampened, should be replaced.)

Next, turn off your flashlight and attic lights and look for pinholes or cracks of daylight in the roof surface. When you have found the source of the leak or leaks, either run a piece of wire or drive a nail through each hole (see sketch at left) to mark the spot for patching from on top of the roof. (Wire will help guide dripping water into a bucket.)

Then, following the instructions that follow for "The emergency patch" and "The temporary shingle," make stopgap repairs.

Examining the roof surface. When the roof is fairly dry again, survey the roof surface. Look for

- the locations of leaks you found when examining the attic
- loose, torn, cracked, or missing shingles or tiles
- moisture and soft wood along the fascia boards, which may have absorbed standing water.

Again, following the instructions given in the following pages, make the type of repair required.

The emergency patch

If you wake up one morning to discover that it's been raining in the breakfast room, locate the source of the leak and plug it temporarily with what's called a "wet surface roof patch."

Roof patch is an asphalt or coal/tar compound, similar in appearance to ordinary roofing cement. It can be applied to both wet and dry surfaces, either from the attic or roof surface.

You can find roof patch in one-gallon cans at home improvement centers, hardware stores, and roofing supply companies.

Using a caulking gun and putty knife, simply apply roof patch liberally to the leak. Then rub the spot thoroughly so that the compound will adhere properly.

The temporary shingle

If the source of a leak lies among damaged or missing shingles, first plug the leak with roof patch on the underside. When the roof is dry enough to be accessible, slip a 2-foot square of sheet metal under the row of shingles above the defect (see sketch below). The sheet metal will

The temporary shingle

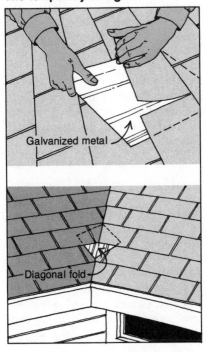

Galvanized metal

Diagonal fold

help carry water away from the leak until weather conditions permit a more permanent solution.

If the leak lies along a closed valley, use an 8-inch square of sheet metal folded diagonally to fit the angle of the valley. Slide it under the shingles over the defective area.

Permanent solutions

Using ordinary plastic roofing cement and roofing nails, you can often permanently plug a leak that's been traced to a crack or tear in asphalt shingles or wood shingles or shakes (see "Simple Repairs: Asphalt & Wood," following).

You'll need to do a little more work to repair a leak where the caulking around flashing has shrunk or cracked, where metal flashings have corroded, or where shingles or tiles have been torn or blown away (see pages 74–76).

Where a long-ignored leak has done structural damage, it may be necessary to replace sections of sheathing or bolster weakened rafters (see page 77).

SIMPLE REPAIRS: ASPHALT & WOOD

Use plastic cement and, if necessary, roofing nails to salvage asphalt shingles or wood shingles or shakes that have split or torn.

Repairs for asphalt shingles

When you repair asphalt shingles, pick a day when the roof is dry enough to be accessible and the air is warm enough to make the shingles pliable. In addition, have your roofing cement at room temperature before you use it, so that it will spread easily.

Small tears. To repair a small tear in an asphalt shingle, use a putty knife to apply cement under the defective area. Then press the shingle back into place.

For hairline cracks, use asphalt paint to seal the defect.

Large tears. To repair a large tear in an asphalt shingle, trowel plastic cement liberally underneath, press the shingle back into place, and secure it with a nail on either side of the tear (see drawing, next page). Then daub cement over the nailheads.

Repairing asphalt shingles

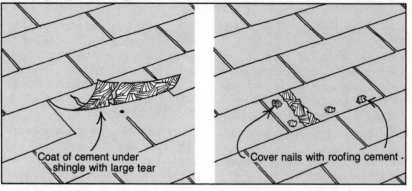

Coat of cement under shingle with large tear

Cover nails with roofing cement

Curled shingles. To flatten curled shingles, apply cement under the lifted portions. Press them back down and tack in place with roofing nails. Cement over the nailheads.

Repairs for wood shingles or shakes

Use roofing nails and dabs of plastic cement to repair a splintered or cracked shingle, or one that has been lifted by the wind.

Split shingles. To repair a split or cracked shingle, cut or pull away the splinters so that only large, solid pieces remain. Butt the pieces tightly together and, after drilling pilot holes so that they won't split, nail each piece in place (see sketch below).

Repairing wood shingles

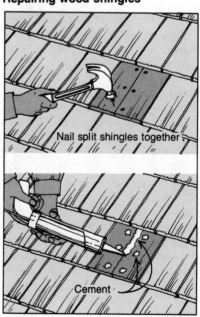

Nail split shingles together

Cement

Then cover the nailheads and fill the joints with cement between pieces.

Wind-lifted shingles. Wind-lifted shingles need only be pressed down flat and secured with a nail. Daub plastic cement over the nailhead to provide a seal.

REPLACING FAULTY SHINGLES & TILES

To replace worn shingles or broken tiles, you will need a pry bar, a hacksaw blade, a hammer, and—for wood and tile roofs—a shingle ripper (available through roofing suppliers).

To use the shingle ripper, slide it underneath the shingle and around each nail; then cut the shank of the nail with a hammer blow.

Replacing asphalt shingles

Where you've discovered badly worn or damaged asphalt shingles, replace them if possible with some that remain from when the roof was originally installed.

If you don't have leftover shingles on hand, buy a bundle of new shingles—identical in brand, color, and size, if possible—to use as replacements.

How to remove shingles. To remove an asphalt shingle, carefully lift up the shingle tab above the defective shingle, as the illustration at right indicates. Then, using a pry bar, remove the lower row of nails from the shingle you want to replace.

Using the pry bar again, pull the second row of nails—under the tabs of the shingles two courses above—to free the defective shingle.

Replacing the shingle. When replacing a lone shingle, just slip it into place, taking care not to damage the roofing felt. (Snip off the upper corners if the shingle sticks.)

Nail down the new shingle with two new sets of nails in the old locations. Apply daubs of cement to the nailheads.

Replacing asphalt shingles

1. Pry out both sets of nails and remove damaged shingle.

2. Slide new shingle into place and nail.

3. Use a screwdriver to offset hammer when nailing.

If you've removed more than one shingle, treat the new ones as if you were laying whole courses. Work from the eave to the ridge (see pages 65–67), and install the last shingle in the manner described here.

How to replace wood shingles and shakes

1. Split and remove damaged shingle; cut off nails with a shingle ripper or hacksaw blade.

2. Tap new shingle into position.

3. Nail as shown, then cover nailheads with cement

Replacing wood shingles or shakes

Replace shingles or shakes that have been badly splintered or curled, as well as those that have begun to crumble.

Removing worn shingles or shakes. Splitting the shingle along the grain, pull out as much of the damaged shingle as possible. Then pry up the shingle or shingles directly above it—far enough to slip a hacksaw blade or shingle ripper under it (see the sketch above). Cut or saw the nails off flush with the sheathing, but be careful not to damage the sheathing or underlayment in the process.

Cut through all four nails that secure the shingle—the two just above the butt line of the course just above and two more above the butt line of the next higher course.

Replacing shingles or shakes. Use a roofer's hatchet to trim a shingle or shake to suit the width of the piece you've removed. Then, taking care not to damage any felt roofing paper, slide or tap the new shingle or shake into place. If you use a hammer, place a short length of 1 by 2 between it and the new shingle so that you don't chip the new material as you install it.

Secure the new shingle with four nails, placed 1 inch from its vertical edges. Drive two of them ½ inch above the butt line of the course just above, and two more ½ inch below the butt line of the next higher course. Daub roofing cement over the nail heads.

Replacing tile

If you're replacing cracked tiles on a roof where tiles have been laid over the surface, the job is a simple one: lift up the tile or tiles in the course above the cracked tile, remove the old tile, and slide in a new one.

If you're replacing tiles that have been nailed to furring strips, however, you will need a shingle ripper or hack saw to cut the nails before you remove the broken tiles.

First use a hammer to break up the tile into pieces. Then remove as much of the tile as you can.

Next, use a pry bar to carefully lift up the tile or tiles directly above the broken one, and remove the remaining tile shards and nails with a shingle ripper or hack saw. Fit a new tile into position, but do not nail it.

HOW TO PATCH A BUILT-UP ROOF

If you've tracked down a leak to a sizable bump or split in your built-up roof, a little minor surgery should return the roof to normal.

Repairing a blister

To repair a blister in a built-up roof, first sweep back the gravel surface with a stiff-bristled broom. Then use a utility knife to cut into the asphalt and roofing paper until the pressure under the blister is released.

Next, use a putty knife to cover the

cut and an area 2 inches beyond each side of it with plastic cement, working the cement well into the cut.

From 30-pound roofing felt or asphalt shingles, cut a rectangular patch 2 inches larger than the slit in all directions. Press the patch into the cement, nail around its outer

Repairing a blister

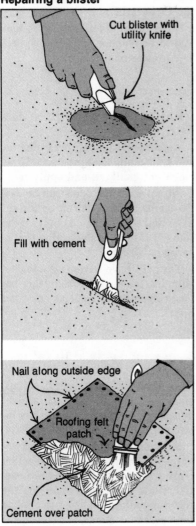

Cut blister with utility knife

Fill with cement

Nail along outside edge

Roofing felt patch

Cement over patch

edges, and then cover the patch with another layer of cement. When the cement has begun to dry, cover the patch with a layer of gravel.

Plugging holes

To patch a hole (up to one square foot), sweep gravel aside, cut out a rectangle of the damaged layers of roofing, and carefully remove them.

Then fill the rectangular hole with roofing cement and cover it with a patch cut from either 30-pound roofing felt or asphalt shingles. Nail the patch along its outer edges and cover it with liberal amounts of roofing cement. Extend the cement 2 inches beyond each edge of the patch.

Cut a second patch 2 inches larger on each side than the first. Nail it in place, cover it with plastic cement, and replace the gravel.

Holes larger than one square foot should be patched with hot asphalt by a professional roofer.

REPAIRING FLASHINGS

To repair leaks that spring from broken seals along flashings, simply remove old caulking and apply new.

Pinholes or rust spots can be easily patched with a dab of roofing cement and a small patch cut from the same material as the flashing that is being repaired.

You will have to remove adjacent shingles to repair broken seals beneath flashings or replace corroded metal.

Renewing seals at flashings

To restore visible seals at flashings, simply chisel away the old sealant and apply new according to the instructions that follow.

If, though, you've found a leak *underneath* flashings, you must either work fresh sealant under the material with a putty knife or remove the flashing entirely in order to apply a new coat of cement.

Restoring visible seals. To restore visible seals that have aged and cracked, first chip away old sealant with a hammer and chisel. Then, after roughening the surface with a stiff wire brush or coarse sandpaper, use a caulking gun filled with roofing cement (not roof patch compound) to apply new seals.

Restoring seals at flashings

Chip old sealants away.

Apply new sealant.

To ensure a tight seal, use cement that has been kept at room temperature, particularly if outdoor temperatures are below 50°F/10°C.

Restoring seals under flashings. Using a caulking gun and a putty knife, inject enough roofing cement to fill the trouble spot, then surround the area with a bead of cement.

If you must remove flashing in order to get at the defective seal, first remove adjacent shingles according to instructions described under "Replacing Faulty Shingles & Tiles," page 74. Then carefully pry flashings away from their seals, using the flat edge of a putty knife for leverage. (Be sure to wear gloves when working with metal.)

Once you've removed the flashing, chisel away old sealant and resurface the area with a liberal coat of fresh cement. Put the flashing back in place, embedding it in the new cement, and replace surrounding shingles according to the instructions on pages 74–75.

Patching holes in flashings

You can plug small holes in flashings with spots of plastic cement for holes the size of pinheads) or with patches (for holes up to the size of dimes). Flashing with larger holes should be replaced.

To patch a hole in flashing, use the same material as the flashing, cutting a square that extends 2 inches beyond the hole on all sides.

Next, use a wire brush or coarse sandpaper to roughen the area around the damaged part of the flashing. Remove grit with a clean, damp cloth. Daub plastic cement on the spot, and press the patch firmly into the cement. Hold it in place for several minutes, then cover the patch with another coat of cement.

Repainting metal flashings

Prepare the surface of any metal flashings you are repainting with a stiff brush and solvent that will remove loose paint, rust, and corrosion. (Take care not to spill solvent on asphalt shingles.)

Then, after applying a zinc-base primer, spray on two or more light coats of rust-resistant metal paint. To avoid getting paint on the roofing materials, tape newspaper to the roof around the flashing before you begin.

Replacing flashings

If the defective flashing has been corroded or weakened beyond repair, remove adjacent shingles according to the instructions under "Replacing Faulty Shingles & Tiles," install new flashing (see pages 57–62), and replace the shingles.

REPAIRS FOR SHEATHING & RAFTERS

Unfortunately, it happens now and then that dry rot—actually an advanced form of mildew—gets into the sheathing and rafters and must be corrected. Rafters themselves, after many years of exposure to alternating dry and moist air, can crack and sag in the middle.

How to replace damaged sheathing

To repair a section of rotted sheathing, remove roofing materials in and adjacent to the damaged area, following instructions for removing shingles or tiles on pages 74–75.

Then, using a circular saw (set the blade to the thickness of the sheathing), cut through the damaged sheathing back to the nearest rafter, as the sketch at right indicates. If the damage is considerable, remove the entire affected panel. (When removing a section of rotted plywood, make cuts over the middle of a rafter; then you can use the rafter as a nailing base for the new material.)

Using the damaged sheathing as a pattern, cut a new piece of plywood (which should be the same thickness as the material it is replacing), and fit the new piece to the roof.

Nail down the plywood, using 2-inch common or box nails for ½-inch plywood or 2½-inch nails for plywood ⅝ to ⅞ inch thick. Nail every 6 inches along the vertical edges of each piece and every 12 inches where sheathing crosses a rafter.

Before replacing shingles, patch resin pockets and knotholes and apply new roofing felt according to instructions on pages 52–55.

Making rafter repairs

Now and then rafters become victim to moisture or dry rot in the wood, or they sag and crack due to years of exposure to damp and then dry air.

Removing damaged sheathing

Use circular saw to cut through damaged sheathing.

Installing 'sister' rafters

Strengthen weakened rafters with 'sisters,' one nailed to each side.

Correcting for moisture and dry rot. Where rafters have absorbed excessive moisture, either because of inadequate ventilation or a leak, take steps to improve ventilation or patch the leak. Then, when the wood is dry, apply wood preservative to protect it from further damage.

Where the wood is affected by rot, you may be able to correct the problem with a coat of good fungicide.

To repair more severely damaged wood, use a knife to dig out as much of the spongy matter as possible. Then fill the hole with a plastic putty compound, which can be applied with a putty knife. (You can find plastic putty at marine supply stores, if not at home improvement centers.) When hardened, the pastic lasts longer than wood because it is invulnerable to decay.

Bolstering a sagging rafter. To strengthen a rafter that is sagging or bowed, you can sandwich it between two supporting "sister" rafters that are nailed to either side.

Using well-seasoned lumber the same width as the existing rafter, cut two pieces long enough to extend 2 feet beyond the damaged area at both ends (see sketch above).

If the damage is near the ridge, miter the ends of the sisters so they fit snugly against and can be toe-nailed to the ridge beam. If the damage is near the eave, cut the ends to fit the sill plate of the house wall.

Then fasten the first sister to the damaged area, using nails long enough to penetrate both rafters; nail every 12 inches. Hammer down exposed points on the other side of the rafter, and install the second sister.

If you want to bring more daylight into a dull, dim room, take a lesson from the artist who works in a free flow of natural light. Even on cloudy days, overhead light flooding through skylights brightens an artist's studio all day long. Because of its position on the roof, even a small skylight lets in three to five times as much light as a conventional window the same size.

Designed for installation by the do-it-yourselfer experienced in basic carpentry, a skylight usually consists of a preframed plastic "window" body you buy and attach to the roof surface. It may also include some kind of shutter or light diffuser added at ceiling level for light control.

CHOOSING A SKYLIGHT

You can choose from a wide assortment of rectangular, square, triangular—even round—skylights that are manufactured to fit standard 16 or 24-inch rafter spacings in your roof (see illustration below). Most have acrylic plastic windows that are either clear and colorless (for maximum light

Ready-to-install skylights

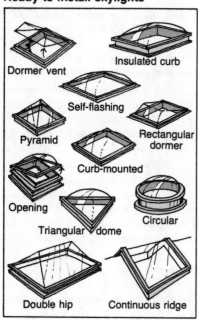

Dormer vent

Insulated curb

Self-flashing

Pyramid

Rectangular dormer

Curb-mounted

Opening

Circular

Triangular dome

Double hip

Continuous ridge

transmission and solar heat gain), translucent white (for soft, diffused lighting), or tinted neutral gray or another color to reduce light and heat without blocking the skyward view.

How large a skylight you should install will depend on the number and size of windows already in the room, and on the function you want the skylight to serve. Will it amplify light in a high-task room such as the kitchen or workshop? Or will it spotlight an indoor garden or brighten a dark hallway?

If you're installing a skylight where the sun is intense, you can avoid hot spots on carpets or furniture by using tinted skylights, built-in shutters, horizontal curtains, or a translucent panel installed below the light shaft.

Double-glazed skylights are available; they prevent heat loss as well as condensation on the inside of the skylight.

INSTALLING A SKYLIGHT

If you're familiar with basic carpentry, installing a square or rectangular skylight on a roof that's moderately sloped (to 6 in 12) and covered with asphalt or wood should present no major problems. Great care must be taken, though, in sealing the opening in the roof and reframing any openings cut in roof rafters and ceiling joists.

Work of this type is always governed by local codes, so be sure to check with your building department first. You may need a building permit.

You'd be wise to call on a contractor to install a skylight larger than 48 inches square, or any skylight on a steeply sloped roof, or on a roof surfaced with slippery tiles or slate.

Tools, materials, and installation tips

Tools and materials you'll need for the project include a hammer, circular or saber saw, straightedge, combina-

tion square, galvanized roofing nails, 2 by 4 or 2 by 6 pieces of lumber for the curb and framing, roofing cement, metal for flashing, and roofing felt.

Step 1: Preparing the opening

Working from the attic or crawl space, lay out the opening you plan to cut—to correspond with the inner dimensions of the skylight—by driving four nails at its "corners" (see sketch below), two each along the inner edges of the rafters to be spanned. Add 3 inches at both the upper and lower edges of the projected opening to allow for rough framing.

Then go out onto the roof and peel back or remove roofing materials (except built-up roofing) 12 to 16 inches beyond the rough perimeter of the opening you plan to cut (see pages 74–75). Save the roofing materials for reapplication once the skylight is installed.

After removing the roofing materials down to the felt, draw a square or rectangle with a straightedge, using the nails as your corner guides. (To be sure you've laid out a perfect square or rectangle, measure diagonally from nail to nail; the distances should be identical.)

Preparing the opening

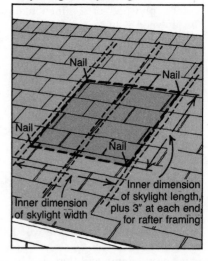

Nail

Nail

Nail

Nail

Inner dimension of skylight width

Inner dimension of skylight length, plus 3" at each end for rafter framing

With a circular or saber saw, cut through the sheathing along the lines. If it's necessary to saw through a rafter or two, make the cuts at right angles to the roof sheathing (use a combination square to mark the rafters).

Step 2: Framing the opening

Where you have cut through a rafter, brace it with double headers the same dimension as the rafter, as the illustration below shows.

Once the headers are installed, cover them with 3-inch-wide strips of sheathing and 6-inch-wide strips of roofing felt.

Step 3: Building a curb

Using well-seasoned 2 by 4 or 2 by 6 lumber, build a curb, or box frame, for the skylight. The curb's inside dimensions should equal the dimensions of the roof opening so that the curb lies flush with the framing.

Next, apply a generous coat of roofing cement around the hole, embed the curb into it, and toenail the curb to the roof.

With the curb installed, reapply roofing materials along the lower edge of the skylight opening so that flashing can be installed on top of the shingles.

Step 4: Flashing the curb

Here's the most critical part of the installation—critical because the role of flashing is to form a watertight seal, and to do this it must be properly applied.

Adapting the pattern and instructions given for chimney flashing on pages 59–60, install metal flashing around all four sides of the curb (see sketch below). Flashing should be embedded in a generous base of roofing cement, then fastened to the curb with galvanized nails. Step flashing along the sloping sides of the curb, embedded in cement on top of roofing felt, should be installed with each course of shingles or shakes.

Finally, reapply roofing materials to within ½ inch of the base of the curb on its upper and sloping sides.

Step 5: Positioning the skylight

Following manufacturer's instructions, install the preassembled sky-light in the curb, using adequate amounts of sealant between curb and skylight.

INTERIOR FINISHING

To finish the skylight, either cover the frame with wallboard (for rooms with finished ceilings) or build a light shaft between ceiling and skylight. You can build a straight shaft like the one illustrated, or you can angle it to adjust the amount of light that will enter the room.

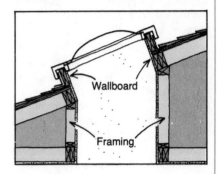

If you're building a light shaft, you'll need to cut a hole in the ceiling in line with the skylight opening; frame the corners of the ceiling opening to support wallboard; line the shaft with ½-inch wallboard; and finish with paint.

Framing the opening

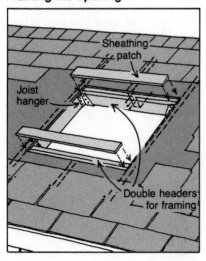

Building the curb

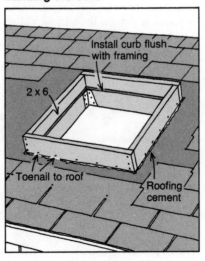

Flashing the curb

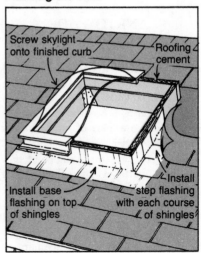

GUTTERS AND DOWN-SPOUTS

They are not so conspicuous, of course, but your house's gutters and downspouts channel water in the same way the great Roman aqueducts did. Both were engineered to move water from one location to another—and, in the case of your own house, from the roof to the ground.

How efficiently your gutter system functions depends on the quality of its installation and maintenance. Improperly sloped, gutters may not be able to move water properly toward the downspouts. Infrequently maintained, they can become clogged with debris that makes them unable to function.

In this chapter you'll find information to help you maintain, repair, and upgrade—or completely replace—the gutters and downspouts around your house.

ANATOMY OF THE SYSTEM

The working parts in a gutter and downspout system (see sketch at right) fit together like pieces in a puzzle that can be arranged to conform to the angles of your house's roof and walls.

You can buy any number of preformed parts—U-shaped troughs, corrugated elbows, and rectangular or round downspouts—to repair damages in an existing gutter and downspout system, or you can buy an entire system for replacement or new construction.

Materials

Most gutters and downspouts are made of galvanized steel, aluminum, or vinyl. In some instances you'll see gutters made from copper and wood.

Galvanized steel. Gutters and downspouts made from galvanized steel are available either unfinished or painted. The least expensive alternative, galvanized steel gutters are strong and heavy. If left unpainted, however, galvanized steel deteriorates rapidly; factory-applied baked enamel finishes offer greater durability. Though they are more expensive, heavy-gauge steel gutters last longer than standard-gauge.

Aluminum. Aluminum gutters and downspouts also come either unfinished or painted, though systems with factory-applied baked enamels are more durable. Aluminum gutters and downspouts are lightweight, easy to work with, and more weather resistant than steel; but their plasticity makes them a little more vulnerable than steel to denting and bending. Like gutters made of galvanized steel, aluminum gutters come in both standard and heavy-gauge metal.

Vinyl. Vinyl gutters and downspouts are more expensive than steel or aluminum, but they wear extremely well with minimum maintenance. Vinyl gutters won't rust, rot, or blister, and they never need painting. Maintenance is simply a matter of periodic cleaning.

One of vinyl's two drawbacks is that it is available only in white or gray.

You can compensate for its other drawback—considerable expansion and contraction at joints because of temperature changes—by installing expansion joints between lengths.

Hangers

Gutters are attached to the house with one of several types of hangers: spike-and-ferrule, strap, and bracket.

Spike-and-ferrule hangers are the simplest and most commonly used

Anatomy of a gutter system

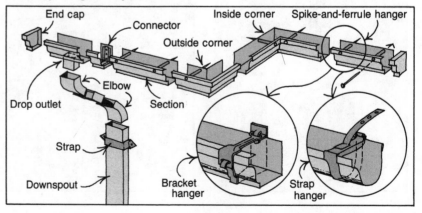

End cap · Connector · Inside corner · Spike-and-ferrule hanger · Outside corner · Drop outlet · Elbow · Section · Strap · Downspout · Bracket hanger · Strap hanger

hangers, though they give out more quickly under the weight of snow and ice. Spikes are nailed through holes drilled in the gutters (and through a sleeve) into the fascia boards.

Strap hangers are nailed onto the roof underneath roofing materials. They are usually installed before starter courses in new wood or tile roofs. They can be fastened to an asphalt roof after shingles have been laid, because bottom tabs can be lifted to install hangers.

Bracket hangers, nailed to the fascia board, support the gutter either with an internal clip or an external band.

INSPECTING GUTTERS

When you examine the gutters and downspouts around the house, take a thin screwdriver and scraping knife along to test for weaknesses in the metal and fascia boards. Then look for

- debris that may have clogged gutters and downspouts
  - flaking or peeling paint
  - cracks in connecting seams
  - rust spots and holes in metal
  - gutters that have sagged due to loose fastenings
  - loosened straps along downspouts
  - soft spots in fascia boards that indicate dry rot.

Use a level to check the slope of gutters. They should tilt away from valleys and toward downspouts 1 inch every 20 feet.

REPAIRS FOR GUTTERS & DOWNSPOUTS

Reviving what is basically a sound roof drainage system may involve nothing more than unplugging downspouts that have been clogged with debris, or resealing a few joints in the gutters.

If you need to correct for dry rot

in the fascia boards, make fascia repairs first. It may be necessary to remove a section or two of gutter to get at the damaged wood.

Unclogging gutters & downspouts

Remove leaves, twigs, and other debris from gutter troughs. (Use gloves.) Then, with a stiff brush and garden hose, loosen dirt that has caked along the inside of the gutters.

If a downspout is blocked, first try to flush debris out with the garden hose turned on full force. If that fails to clear the downspout, feed a plumber's snake into it. Then flush loosened debris out with a hose.

To prevent new litter from accumulating, lay mesh screens over the troughs in the manner illustrated below. The screens will deflect leaves, twigs, and other debris over the edge of the gutter.

You can also protect the downspouts with strainers designed to

Two ways to deflect debris

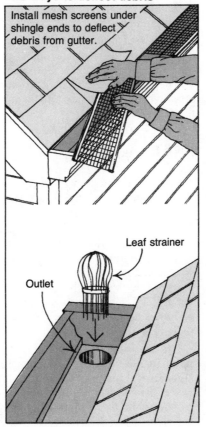

Install mesh screens under shingle ends to deflect debris from gutter.

Leaf strainer

Outlet

admit water while filtering out debris.

Patching leaks

You can seal broken seams or small holes produced by rust in gutters and downspouts with dabs of plastic cement, mastic, or butyl gutter and lap seal.

To repair holes larger than ¼ inch, however, you'll need to cut and apply patches of canvas or aluminum.

Patching cracks and small holes. If a leak develops at a joint, apply silicone sealant around the seam on the outside of the gutter.

To patch a hole smaller than ¼ inch, first brush away dirt from the area and then roughen it with a wire brush or coarse sandpaper (see illustration on next page). Wipe away grit.

Next, apply a thin coat of plastic cement or butyl gutter and lap seal; extend the cement 6 inches beyond the hole in both directions.

Patching large holes. Clean and roughen the area around the defect according to the preceding instructions, and apply a thick coat of plastic cement.

Then embed a patch of aluminum, canvas, or fiberglass mat—cut to extend beyond the hole at least ½ inch on all sides—into the cement and apply another coat of cement over the patch.

Sections of gutter with extensive damage are probably easier to replace than to repair (see "Replacing damaged gutter sections," next page). But you can correct such damage with a patch cut from the same material and cemented in place.

Adjusting gutters

The spikes and straps that support gutters will sag with time, especially in snow country. Every time you clean out accumulated debris, finish by pouring a pail of water into each

Use wire brush to clean damaged area.

Apply roofing cement to 6" area on either side of damage.

For holes larger than ¼", patch with sheet metal and a second coat of cement.

gutter. If the water doesn't drain quickly and efficiently, adjust the bend of the hanger straps to lower the gutter at its downspout and raise the opposite end; the gutter should slope 1 inch for every 20 feet.

Replacing broken hangers

With visegrip pliers, pull out the nail of a defective spike-and-ferrule hanger. Lift shingles to remove broken strap hangers. Use the defective hanger as a guide when buying a replacement.

To avoid weakened nail or screw holes, fasten replacement straps slightly to the side of where the original straps were attached.

Replacing damaged gutter sections

To replace a section of damaged gutter, first lay a scrap of 2 by 4 in the trough to brace the gutter. Then cut out the damaged length with a hacksaw.

Have your building supplier cut a new section—it should be the length of the old section minus the space required for two connectors. (Take the damaged section with you to the supplier for a proper match both in length and gutter shape.)

Install the new piece, sealing the connecting joints with caulking compound in the manner illustrated on page 83.

Correcting for dry rot in fascia boards

If in poking around with your screwdriver you have found soft spots in the fascia boards, either carve out the dry-rotted spot and fill it with a commercially available plastic putty, or remove the damaged section of board entirely and replace it with a new one.

In either case it may be necessary to remove a section of the gutter system temporarily.

When replacing part of a fascia board, use well-seasoned lumber of the same dimensions as the board you're removing. Apply a wood preservative and then paint to match the rest of the fascia.

Repainting gutters

A fresh coat of paint every few years will extend the life of metal gutters, especially where they are exposed to salt air.

First prepare the metal by thoroughly cleaning gutter troughs and wire brushing or sanding down areas of rust or corrosion. Also sand down the gutter exterior wherever the old paint has begun to chip or peel, and seal areas of potential leaks.

Use an asphalt-aluminum gutter paint inside the gutter; on the outside, use a rust-preventative zinc-based primer before you apply an exterior paint that's compatible with the color of the house.

REPLACING GUTTERS & DOWNSPOUTS

Check with several building suppliers and collect manufacturers' brochures to compare the gutter systems available in your area. Compare for cost and durability, as well as ease of installation and maintenance requirements.

Then, consulting brochures and using your existing system as a guide, estimate what you'll need to install a new system for your roof. (Before you install the new system, repair and repaint damaged or weathered fascia boards.)

Estimating materials

There are only a few rules of thumb for estimating materials for a gutter system.

Sizes. Gutters come in 4, 5, and 6-inch diameters; downspouts are available in 3 and 4-inch diameters.

What size you should install depends on the square footage of your roof. Use the following table as a guide.

Roof Area (Square feet)	Gutter Diameter	Down-spout Diameter
100–800	4″	3″
800–1000	5″	3″
1000–1400	5″	4″
1400+	6″	4″

Lengths. You'll find that most manufacturers sell gutters in 10-foot lengths (some are longer); estimate the number of lengths you'll need based on the length of the eaves.

Also count the number of fasteners and other fittings required. Remember that you'll need only one extension joint where two gutter lengths meet. You'll also need caps for the outside corners, and one hanger for every 3 feet of gutter length. Also count the number of miter joints necessary, and calculate the length and number of the downspouts.

Downspouts. Allow one downspout for every 20 feet of gutter. Install downspouts at both ends of a gutter if the eave is longer than 35 feet (see illustration below).

Assembling & installing the system

Assembling and installing a gutter and downspout system is most easily accomplished if you have a helper working with you. Together you can assemble gutter runs on the ground, and then fasten them along chalk lines to the fascia boards. (See illustration below.)

Installing a gutter system

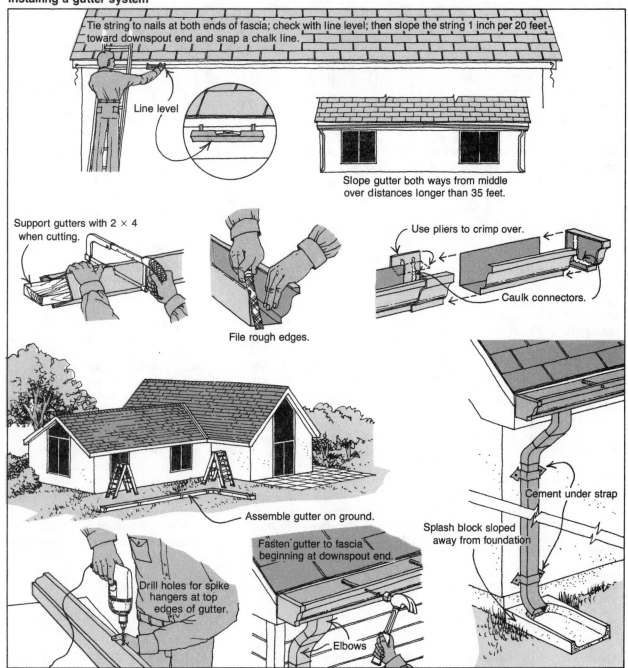

Tie string to nails at both ends of fascia; check with line level; then slope the string 1 inch per 20 feet toward downspout end and snap a chalk line.

Line level

Slope gutter both ways from middle over distances longer than 35 feet.

Support gutters with 2 × 4 when cutting.

File rough edges.

Use pliers to crimp over.

Caulk connectors.

Assemble gutter on ground.

Drill holes for spike hangers at top edges of gutter.

Fasten gutter to fascia beginning at downspout end.

Elbows

Cement under strap

Splash block sloped away from foundation

Snapping a chalk line. In order to provide the proper slope for a gutter system—1 inch for every 20 linear feet—snap a chalk line along the fascia board.

Position the line immediately below the roofing material where the gutter will be located, and tie it to nails at each end. Use a line level to be certain the string is level (don't trust a visual estimate).

Then lower the nail at the downspout end the appropriate distance to achieve a 1-inch drop for every 20 feet of gutter. Snap the chalk line and install the gutter against it.

Assembling the parts. Gutter components either snap together or are joined with connectors. Some vinyl systems are glued.

If the gutter assembly has connectors, apply a sealant to the inside of the connector and the underside of the gutter sections as illustrated. Then push the ends together. Use pliers to crimp the end of the connector over the edge of the gutter. Also caulk end caps.

If you are installing a vinyl system whose parts are glued together, don't glue the downspouts in place; it's not necessary, and you may want to remove them later on when you're clearing the gutter system of debris.

How to cut gutters. Whenever you cut gutter pieces to fit, use a 2 by 4 as a support and a hacksaw to cut through the material.

Use a file to sand down burrs and rough edges so connecting pieces fit properly.

Fastening the gutters. If you have a helper and a couple of ladders, installation will go quickly. If not, tie a wire around one end of the gutter and hook it over a nail driven temporarily into the eave.

Nail the first hangers at the downspout end to position the outlet section accurately over a splashblock. Use two hangers at each corner for adequate support.

Assembling and connecting downspouts. Downspout components fit one inside the other and may be joined to the gutter assembly with elbows on a house with overhangs.

On metal downspouts, drill two holes for metal screws to secure the downspout or elbow at the drop outlet.

To secure downspouts to the house siding, nail or screw straps into wood; use a masonry bit to drill holes for expansion bolts in stucco; in brick walls, drill holes with a masonry bit and drive in lead anchors for the screws.

Coat the back of each strap with caulking compound or roofing cement to produce a seal.

Testing the system. Once you have the entire gutter/downspout system installed, turn on the garden hose and run water from the end of each gutter toward the downspout. Watch carefully for leaks or insecure joints that may weep water; then make whatever adjustments or repairs are necessary.

Splashblocks & dry wells

Water that flows from your downspout directly to the ground eventually ends up either in your foundation or your basement. And it erodes the soil alongside the house.

Water is ordinarily diverted from the house either by splashblocks located at the foot of downspouts, or underground drainage pipes that carry water to a dry well several yards from the house.

Splashblocks. Ready-made concrete splashblocks, placed below an elbow attached to the downspout, should be tilted slightly so water flows outward.

An alternative to splashblocks is a plastic or fabric sleeve, available at home improvement centers, that is attached directly to the downspout. Holes in the sleeve allow water to seep out slowly over the ground.

Dry wells. If you live in a wet climate, you might want to link the downspout to a dry well located 10 feet or more from the foundation.

To make a dry well it's necessary to dig a hole large enough to accommodate a 55-gallon oil drum, and a trench deep enough to accommodate drainage pipes that are sloped about ½ inch per foot from the downspout to the well (see sketch above).

The drum, punctured with holes and filled with either rocks or concrete blocks, should be located at least 10 feet from the foundation and at least 18 inches below the ground.

Typical dry well

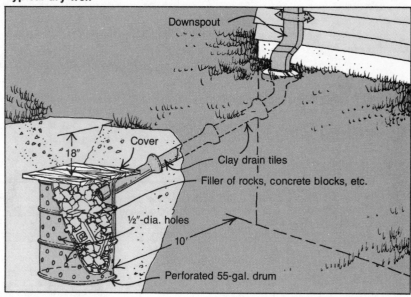

Downspout

Cover

18″

½″-dia. holes

10′

Perforated 55-gal. drum

Clay drain tiles

Filler of rocks, concrete blocks, etc.

SIDING

SELECTION, CARE & INSTALLATION

Where the preservation of your house is concerned, siding is every bit as important as a good roof. Just step outside in 100-degree heat or in a wind-driven rain or snow storm, and you'll know what your house exterior has to contend with.

But siding is more than protection. The color, texture, materials, and design of siding make up the exterior grooming of your house—and the choices available are myriad, with infinite pros and cons for every type and category.

It's perhaps one homeowner in fifty who totally replaces house siding, but every homeowner must deal with maintenance and repairs. And for one who is building a new house or adding on, the choice of siding material is a design decision almost as important as the floor plan.

Whatever siding material you choose, you'll be making a major investment. You'll also be surrounding yourself with a sizeable surface that you—and your neighbors—will be looking at every day for years, so you'll want to be sure the siding wears well from an esthetic, as well as a practical, point of view.

This chapter offers you a closer look at the many different kinds of siding. They are compared in the chart on pages 88–89, where information on the characteristics and relative merits of eight wood and synthetic sidings are set forth. And on pages 110–111 is a chart that details the various kinds of wood board siding.

In addition, this chapter will help you diagnose problems that may affect your present siding, and prescribe remedies. You'll learn how to handle many different siding tasks yourself—from maintenance to residing—and how to get professional help when the job is too big.

THE ANATOMY OF A WALL

If you've never peeled open a wall, you may be wondering what mysteries lie behind the wall coverings. Most wood-frame walls are very simply constructed, with 2 by 4 wall studs as the primary framework, as shown in the drawing below.

Wall anatomy

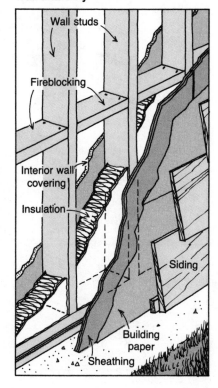

Wall studs

Fireblocking

Interior wall covering

Insulation

Siding

Building paper

Sheathing

Studs are often (but not always) covered with a layer of panel or wood-board sheathing on the outer side. Sheathing gives a wall lateral strength and provides a flat base for nailing on the siding. For more about sheathing, see page 105.

An underlayment of building paper is required between sheathing and most types of siding. Building paper is discussed on page 105.

Flashing helps seal horizontal seams between exterior materials where water would otherwise work its way in. Typically, you'll find Z-shaped flashing at horizontal joints between panels of sheet-type sidings, and drip cap flashing carrying water over the tops of windows, doors, and similar breaks in siding. Flashing is discussed in depth on pages 106–107.

Insulated walls most often have insulation batts or blankets stapled between studs (with the vapor barrier facing inward) or loose insulation blown into the cavities between studs. On some walls insulation is applied in the form of foam in the pockets between studs or, in the form of rigid boards, fastened to the outer sides of wall studs. (For more about insulation, see page 103.)

"Fireblocking" of short 2 by 4s is often added between studs, and "plates" of horizontal 2 by 4s are nailed along the top and bottom ends of studs.

An interior wall covering, such as gypsum wallboard, is fastened to the inside of the wall studs, and some form of siding is applied to the outer faces of the studs. This brings us to our next topic: a closer look at siding materials.

SIDING MATERIALS: A DIZZYING ARRAY

The range of wood-based siding materials includes traditional solid boards, shakes and shingles, hardboard, and exterior plywood.

Add sidings of aluminum, steel, vinyl, and the choices can make you dizzy. (Other favorites, brick, stone, imitation stone and brick, and stucco, aren't discussed in detail, being beyond the scope of easy homeowner installation; stucco, though, is included in the chart on pages 88–89.)

To help sort these out, the chart compares and contrasts the major siding materials available on the market.

Though we don't recommend that you attempt to install all types of siding, we do give a brief rundown on each in the chart. Armed with this knowledge, you can hire a professional to install some of them—such as masonry and stucco—or take on the more manageable materials yourself.

Relative costs are not included in the chart for two reasons. First, if your present siding needs only some fixing up, price will probably not be a major concern; making repairs will cost far less than replacement, however expensive the materials. Second, estimating costs would be an inexact science; prices change constantly, and prices for a particular kind of siding vary widely from region to region.

Savvy buying

Some siding materials—particularly the wood-based ones like boards, shingles, shakes, and plywood—are distinguished in quality by several characteristics. Number of defects, moisture content, species—these are just a few of the considerations that will determine the quality of the materials you buy. Following are some tips for buying those do-it-yourself sidings that need a closer look.

Wood-board siding—a time-honored siding material because of its availability, natural appearance, and

Wood defects, grain

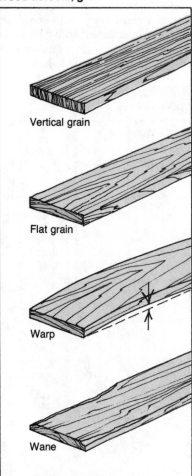

Vertical grain

Flat grain

Warp

Wane

wide variety of styles—is probably the most characteristically American house exterior.

Before buying wood, become familiar with grades. You'll find big variations in price and appearance from one grade to another. The best grades are generally called "Clear" or "Number 1." Lesser grades are "Number 2," "Number 3," and so forth.

Also, if you choose wood siding, you'll probably want it free from knots and pitch pockets; avoid any wood that is warped or has waney edges.

Wood siding is sawn from a log so that it presents either "flat" or "vertical" face grain (see drawing at left). Vertical grain minimizes seasonal movement of the wood due to changing moisture content, and it will hold a painted finish longer; but you'll pay a premium price for it.

Textures available include rough-sawn, smooth, and other variations. You can buy lumber surfaced on any number of sides, though the most common is "S-4-S": surfaced on four sides.

Moisture content is an important consideration. You can buy siding either kiln-dried or "green." Kiln-dried wood has a stabilized moisture content of 12 to 20 percent and won't shrink much. But "green" wood, with a higher moisture content, will shrink. You must allow for this when nailing and joining your siding.

Know the species. Some are better than others for siding, and some species have heartwood that differs from the sapwood. (Heartwood is found near the tree's center, sapwood toward the bark.) Redwood, for instance, has a reddish heartwood that is highly resistant to decay, and thus excellent for siding. The sapwood is whitish and not particularly decay-resistant.

Other species that are excellent for siding include cedar and cypress, followed by some pines.

Finally, be aware that nominal dimensions are generally different from actual dimensions. Boards are given nominal dimensions before drying and milling.

Shingles and shakes. Because these double as roofing materials, they are discussed in depth on pages 44 and 45.

In addition to those made for roofs, you can get a variety of specialty shingles and shakes for walls (see drawing). Or you can buy "sidewall shakes"—actually shingles that have been given deep, machine-grooved

Specialty shingles

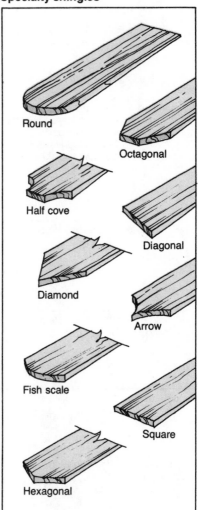

Round

Octagonal

Half cove

Diagonal

Diamond

Arrow

Fish scale

Square

Hexagonal

faces, parallel edges, and straight ends. They are available unfinished or preprimed with paint.

Cedar shingles and shakes are flammable, like all wood products. If this is a problem in your area, see if you can buy them pretreated with a flame retardant.

Plywood siding is divided into grades, depending on the number of patches required to repair the surface in manufacturing. The best has no patches; the worst has no limit of patches. The patches may be of wood, synthetic resin, or both. Of course, if you plan to give the plywood a transparent finish, patches can be blotchy and unsightly. But they are smooth and flat, so if you plan to paint, they shouldn't be a problem. By buying a low grade of plywood, you can save substantially on material costs.

Depending on their rigidity, plywood panels are approved for application over studs that are either 16 or 24 inches apart.

Manufactured sidings. Makers of vinyl, aluminum, and steel sidings have developed a wide range of accessories that complete the siding as a system. Siding panels fit snugly into molded inside and outside corner pieces. Special trim hides the edges of panels below windows and soffits. Trim pieces in the shapes of "J" and "F" have deep grooves for fitting panels around windows and doors.

Among other accessories are starter strips, drip caps, and soffit panels. You can learn more about these in the sections that discuss installation of manufactured sidings.

In addition, some styles of vinyl, aluminum, and steel siding are manufactured with insulation backings or equipped to accept drop-in insulation panels.

More information?

Happily for the buyer, siding manufacturers publish bushels of illustrated brochures and booklets. If

you're tackling a major siding project, the cost of materials can purge your savings account. In order to make the right decisions, it's a good idea—and can be fun—to spend some time collecting available information at lumberyards and home improvement centers.

Of course, when reading brochures, you'll have to regard some claims with polite skepticism. Be wary of phrases like "maintenance free" and "lasts a lifetime." On page 46 of this book, you'll find some information on the warranties offered by roofing manufacturers. Read it—it applies to siding also.

But reading brochures can educate you about both a product and its competitors. Many manufacturers will emphasize their competitors' faults. For instance, makers of synthetic siding often point out that their products won't rot, warp, crack, or delaminate—all of which can happen to unprotected wood products.

Keep in mind that what seems attractive now may not prove to be a benefit later. For instance, though not having to paint vinyl siding for the next 20 years might seem very appealing, you may *want* to change the looks of your house in 5 or 10 years. Vinyl doesn't retain paint well, so you can't give it a simple facelift.

Local contractors are another source of information. Look in your phone book's Yellow Pages under "Contractors—Alterations." Contractors who build room additions keep up to date on most siding products.

Vinyl, aluminum, and steel sidings—as well as other specialized manufactured sidings—are generally distributed through a network of dealer/installers. For information on these, look up "Siding" in your phone book.

Some siding companies have display room mockups of the various products they install. Of course, each siding specialist will tend to "push" the material that he or she sells (or that is most profitable to install), so get as wide a range of impressions as possible.

A COMPARISON OF DO-IT-YOURSELF SIDINGS

Material	Types and Characteristics	Durability
Wood Boards	Available in many species. Redwood and cedar have natural resistance to decay. Milled in a variety of patterns. Depending on type, may be applied horizontally, vertically, or diagonally. For more information on patterns and application, see chart on pages 110–111. Nominal dimensions are 1″ thick, 4″ to 12″ wide, random lengths to 20′. Bevel patterns are slightly thinner; battens may be narrower. Sold untreated, treated with water repellent, primed, painted, or stained.	30 years to life of building, depending on periodic maintenance.
Cedar Shingles and Shakes	Mostly western red cedar; some eastern white cedar. **Shingles** are distinguished by grade: #1 (''Blue label'') are the best; #2 (''Red label'') are second best and quite acceptable as an underlayment when double coursing. Also available in a variety of specialty patterns (see page 87). **Shakes** are available in four shapes and textures, varied by scoring, sawing, or splitting. Also sold in the form of ''sidewall shingles,'' specialty products that are basically shingles with heavily machine-grooved surfaces. Widths are random, from 3″ to 14″. Lengths: 16″ (shingles only), 18″, and 24″. Shakes are thicker than shingles, with butts from ⅜″ to ¾″ thick. For saving labor, 9″ and 18″ shingles and shakes come prebonded on 4′ and 8′ plywood panels. Primarily sold unpainted. Also available prestained or painted.	20 to 40 years, depending on heat, humidity, and maintenance.
Exterior Plywood	Made from thin wood veneers (or ''plies'') laminated under pressure with waterproof glue. Most typical siding species are Douglas fir, western red cedar, southern pine, and redwood. Exterior ply determines designation (interior plies may be mixed). Broad range of textures, from rough or resawn to smooth overlay for painting. Typical pattern has grooves cut vertically to simulate board-and-batten siding. **Sheets** are 4′ wide, 8′ to 10′ long. They are applied either vertically or horizontally. **Lap boards** are 6″ to 12″ wide, 16′ long. Thicknesses of both: ⅜″ to ⅝″. Sold untreated, pretreated with water repellent, primed, painted, or stained.	30 years to life of building, depending on maintenance.
Hardboard	Made from fibers of wood chip pulp, compressed and bonded under heat and pressure. Available smooth or in textures including rough-sawn board, stucco, and many more. **Sheets** are 4′ wide, 8′ to 10′ long. They are usually applied vertically. **Lap boards** are 6″ to 12″ wide, 16′ long. Thicknesses of both: ⅜″ to ½″. Sold primed, primed and painted, or opaque stained.	30 years to life of building, depending on maintenance. Prepainted finish guaranteed to 5 years.
Vinyl	Extruded from polyvinyl chloride (PVC) in white and pastel colors. Smooth and wood-grain textures are typical. Horizontal panels simulate single 6″ and 8″-wide lap boards. Vertical panels simulate single 8″-wide boards with battens. Other styles: single panels that simulate two 4″, 5″, or 6″-wide lap boards, or three 4″-wide lap boards. Standard length: 12′6″.	40 years to life of building.
Aluminum	Extruded panels in a wide range of factory-baked colors; textures. Types and dimensions are the same as vinyl. Also sold as 12″ by 36″ or 48″ panels of simulated cedar shakes.	40 years to life of building.
Steel	Extruded panels in a wide range of factory-baked colors; smooth and wood-grain textures. Types and dimensions are the same as vinyl.	40 years to life of building.
Stucco	Compound made from three parts fine sand, one part Portland cement, and water. Applied wet over wire lath in two or three coats. Pigment added to final coat, or can be painted when dry.	Life of building.

Maintenance	Installation	Merits and Drawbacks
Edges should be treated with water repellent before installation. Needs painting or opaque staining every 4 to 6 years, transparent staining every 3 to 5 years, or finishing with water repellent every 2 years.	Difficulty varies with pattern. Most are manageable with basic carpentry skills and tools.	**Merits:** Natural material. Provides small measure of insulation. Broad range of styles and patterns. Easy to handle and work. Takes a wide range of finishes. **Drawbacks:** Burns. Prone to split, crack, warp, and peel (if painted). Species other than redwood and cedar heartwoods are susceptible to termite damage when in direct contact with soil, and to water rot if not properly finished.
In hot, humid climates, apply fungicide/mildew retardant every 3 years. In dry climates, preserve resiliency with oil finish every 5 years.	Time-consuming because of small pieces, but manageable with basic skills and tools plus a roofer's hatchet.	**Merits:** Rustic look of real wood. Provides small measure of insulation. Easy to handle and work. Easy to repair. Adapts well to rounded walls and intricate architectural styles. **Drawbacks:** Burns. Prone to rot, splinter, crack, and curl; may be pried loose by wind. Changes color with age unless treated.
Before using, seal all edges with water repellent, stain sealer, or exterior house paint primer. Restain or repaint every 5 years.	Sheets go up quickly. Manageable with basic carpentry skills and tools.	**Merits:** Easy to apply. Provides small measure of insulation. Can serve as both sheathing and siding, adding great structural support and strength to a wall. Less expensive than wood boards, yet offers same type of appearance. Broad range of styles and patterns. **Drawbacks:** Burns. May "check" (show small surface cracks) or delaminate from excessive moisture. Susceptible to termite damage when in direct contact with soil, and to water rot if not properly finished.
Before using, seal all edges with water repellent, stain sealer, or exterior house paint primer. Paint or stain unprimed and preprimed hardboard within 60 days of installation; then repaint or restain every 5 years.	Sheets go up quickly. Manageable with basic carpentry skills and tools.	**Merits:** Uniform in appearance, without typical defects of wood. Easy to apply. Numerous surface textures and designs. Takes finishes well. **Drawbacks:** Does not have plywood's strength or nail-holding ability. Susceptible to termite damage when in direct contact with soil, and to water rot and buckling if not properly finished. Cannot take transparent finishes.
None except annual hosing off.	Manageable with basic carpentry skills and tools, plus zipper tool, snap-lock punch, and aviation shears or circular saw.	**Merits:** Won't rot, rust, peel, or blister. Burns, but won't feed flames. Easiest of synthetics to apply and repair. Resists denting. Scratches do not show. **Drawbacks:** Only white and pastel colors available. Doesn't take paint well. Sun may cause long-range fading and deterioration. Brittle when cold.
Needs annual hosing off. Clean surface stains with nonabrasive detergent. Refinish with paint recommended by the manufacturer.	Manageable with basic skills and tools, plus aviation shears or circular saw and brake tool for bending trim. Use aluminum nails only.	**Merits:** Won't rot, rust, or blister. Fireproof and impervious to termites. Lightweight and easy to handle. **Drawbacks:** Dents and scratches easily. May corrode near salt water.
None except annual hosing off. Scratches must be painted to prevent rust.	Manageable with above-average skills. Tools include guillotine cutting tool and brake for bending trim.	**Merits:** Won't rot or blister. Fireproof and impervious to termites. Strong; resists denting. **Drawbacks:** Difficult to handle and cut. Rusts where surface is scratched. May corrode near salt water.
None except annual hosing off. If painted, repaint every 5 years.	Best left to a professional. Repairs are manageable.	**Merits:** Fireproof and durable. Exceptional sheer strength. Provides seamless, even surface. Easy to repair. **Drawbacks:** Difficult and time-consuming to work with, so labor costs are high. Prone to crack as house settles.

SIZING UP THE JOB; ESTIMATING & ORDERING MATERIALS

As labor costs continue to rise, "do-it-yourself" jobs become more and more important for saving money. By doing the work yourself, you can—more often than not—save more than 50 percent on the total cost of the project. But should you tackle the work?

The answer depends on several variables: the scale of the job; the materials involved; the cost of having it done by a professional; your time, temperment, tools, and experience.

SIZING UP THE TASK

If your siding needs only minor repairs and a new coat of paint, you can probably handle the work with little experience. But what about large jobs? Read about the various siding materials, and the techniques for installing them, discussed in this chapter. (Step-by-step methods are not offered in this book for installations that we feel are beyond the abilities of most homeowners.)

If you have the skills but lack a few important tools, consider renting them. Of course, take the cost of rentals into account when you weigh do-it-yourself against professional installation.

Whether or not you decide to do the work yourself, it pays to get bids

from qualified installers. From them you'll get an idea of what you might save by doing your own work, and you may even pick up a few tips on installation.

Is your house exceptionally high? If so, for safety's sake you may want to leave the work to the pros. Is the terrain around your house steep and uneven? It takes experience to rig safe scaffolding on uneven ground. (Many of the safety tips outlined for roofing on pages 40–41 also apply to siding.) Also consider the complexity of your exterior walls. The more corners, angles, windows, doors, and nuances of design, the more difficult the job will be.

FINDING AND ASSESSING A PROFESSIONAL

Look through "Contracting the work out" on page 39. Though it concerns roofing jobs, that information is also applicable to finding and choosing a professional for siding jobs. Two listings to look under in the Yellow Pages are "Siding" and "Contractors: Alterations."

Lumberyards and home-remodeling stores that carry siding materials may be able to recommend installers. And many manufacturers have a network of approved installers for their products. You can get their names by contacting the manufacturers.

Be sure to get competitive bids, and find out how long they'll be valid. Give each bidding contractor the same set of specifications, including details of the work involved, time schedule, completion date, and materials to be used. Be

sure you know who—you or the contractor—will prepare the wall surface and who will haul away scraps and debris.

You might also want to find out whether or not you can participate in some of the work and thus cut down on the cost.

HOW TO WORK OUT SQUARE FOOTAGE

The best way to figure the amount of siding material you'll need is to calculate the square footage of the area to be covered. Gather a pad, pencil, and measuring tape and jot down some fairly accurate measurements.

Divide the surfaces to be covered into rectangles and triangles. Round off measurements to the nearest foot. To get the area of rectangles, multiply length by width. Most triangles will have two sides that join roughly at a right angle. Multiply the length of one of these sides by the other, then divide by two to approximate the area.

You might be able to avoid getting out a ladder to measure the height of a wall if you can measure the width of one board or the length of a shingle exposure and multiply this by the number of boards or courses from top to bottom.

Add together all the areas you've computed, then subtract the areas of windows, doors, chimneys, and other breaks that siding will not cover. The result is the total square footage to be covered with siding. Add 10 percent to this figure for waste. If your house has steep triangles at the gables or other architectural features that will require a lot of cutting, add in another 15 percent.

Plan to keep a small amount of siding on hand for future repairs.

FIGURING FOR TRIM

While you're calculating square footage, estimate the amount of trim you'll need for windows, doors, soffits, corner posts or boards, starter strips, and so forth. These are sold by the linear foot. If you're using wood products, plan to use simple trim moldings in order to keep costs down. Accessory trim products are also available for synthetic sidings. Get a manufacturer's brochure to familiarize yourself with the range of accessories offered.

Trim pieces for synthetic sidings are usually made in two widths: one for use with and one for use without the insulating backer boards that sometimes accompany panels. When ordering, specify the correct thickness for your type of siding.

TIPS FOR ESTIMATING

Depending on the particular type of siding, siding materials are sold by the linear foot, square foot, board foot, or "square" (a roofing term for the amount needed to cover 100 square feet allowing for overlap). Following is a rundown on how to estimate and order each type.

Wood-board sidings are usually sold by the board foot or by the linear foot. Calculate the amount of board feet (1 inch by 12 inches by 12 inches) or linear feet required to cover the necessary square footage. Be sure to take into account the overlap in some styles of wood siding.

Shingle and shake coverage depends on the "exposure" you plan and whether you plan on single or double courses (see page 109). Shingles and shakes are sold by the square. One square of shingles is four bundles; a square of shakes is five.

A square may cover more than 100 square feet of a wall, depending on the exposure. Greater exposures are allowed on walls than on roofs.

The table below will help you determine the actual coverage you'll get with various exposures of shingles and shakes.

Plywood and hardboard sheets are easy to order, because they're sold in sizes that correlate directly with square footage. When ordering, add 10 percent for normal waste and— if your house has sharply angled walls—another 15 percent for extra cutting.

Plywood and hardboard lap panels, because they have an overlap, are not as easy to estimate as sheets. Be sure to increase the amount of material needed, taking into account enough material to allow for overlap.

Vinyl, aluminum, and steel siding materials usually come packaged in cartons with enough material to cover 200 square feet of wall area.

When ordering these synthetic sidings, be sure to estimate the linear feet of various trim pieces you'll need. Consult the manufacturer's brochure, available from your supplier, to see what range of accessories are available for your type of siding.

PROTECTING YOUR MATERIALS

After you purchase and before you install siding materials, they require careful storage. Wood products can be damaged by moisture. For maximum protection, store them in a garage. If that isn't possible, put them up on sawhorses or concrete blocks and planks outside and cover them with a tarp, fastening it down securely. Allow air to circulate from below.

Stack all materials carefully to avoid denting, scratching, or crushing them. Distribute the weight evenly from one layer to the next by separating layers with short lengths of 2 by 4.

Approximate Square-Foot Coverage of One Square of Shingles or Shakes

Exposures:	5½"	7½"	8½"	10"	11½"	14"	16"
Shingle Lengths							
16"	110 sq. ft.	150 sq. ft.	170 sq. ft.	200 sq. ft.	230 sq. ft.	—	—
18"	100 "	136 "	154½ "	181½ "	209 "	—	—
24"	73½ "	100 "	113 "	133 "	153 "	—	—
Shake Lengths and Types							
18" Hand split	55 "	75 "	85 "	100 "	—	—	—
24" Hand split	55 "	75 "	85 "	100 "	115 "	—	—
18" Straight split/true edges	55 "	—	—	—	—	—	112 sq. ft.
18" Straight split	65 "	90 "	100 "	—	—	—	—
24" Straight split	65 "	75 "	85 "	100 "	115 "	—	—

HOW TO MAINTAIN & REPAIR SIDING

With siding, the need for maintenance or repairs is not always as obvious as you may think. Warped siding, shingles flying away in strong winds, crumbling bricks, and peeling paint are hard to miss, of course. But interior dry rot or damage from termite or carpenter ant infestation can easily escape your notice, and these are usually much more threatening to your house.

Obvious or subtle, problems with siding deserve prompt attention. If you neglect them, you'll end up paying for it in the end. Preventive maintenance is the name of the game.

This chapter will show you how to detect typical problems and it will suggest solutions. You'll learn how to maintain, repair, and protect your siding.

MAKING AN INSPECTION

Though you may well want to call in a professional, a preliminary inspection of your own can't hurt. To examine the interior of walls, use a screwdriver to probe for dry rot or insect damage. In some instances, you may need a hammer to give your screwdriver more authority.

You can also use a screwdriver and hammer to examine your house's exterior; but keep in mind that if you gouge your siding and find no fault, you'll have a repair to make anyway.

Pay particular attention to any portion of siding that is in contact with the ground—even indirect contact, as where a fence post abuts the house. Any siding-to-ground contact serves as a conduit for insects and dry rot (though in the case of metal siding, electrical grounding is necessary).

Typical problem areas are illustrated below.

Whether you plan to do the work yourself or contract the job, it's a good idea to seek some professional advice on what needs to be done—especially if you've had no experience sleuthing through crawl spaces (watch your head!), attics, and basements. There are two reasonable ways to accomplish this. One is to ask a local contractor for a bid on the project, explaining that you are thinking of doing it yourself and that you'll pay him an hourly fee for his advice. The other, if you suspect you might have problems with termites or other wood-chewing critters, dry rot (which is caused by a fungus), or wood-boring beetles, is to get an inspection by a licensed termite inspector or pest control professional.

Termite troubles

When extreme, termite damage is obvious, but in most cases it is sub-

Problem areas

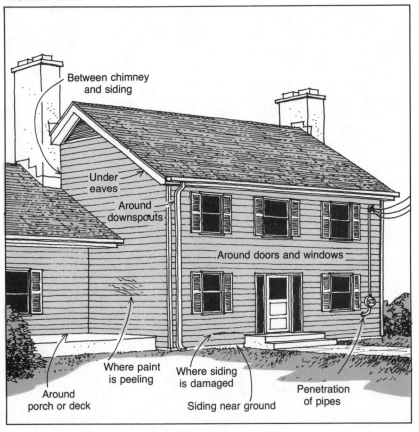

Between chimney and siding

Under eaves

Around downspouts

Around doors and windows

Where paint is peeling

Where siding is damaged

Around porch or deck

Siding near ground

Penetration of pipes

tle until one day the floor caves in or the roof sags.

Dry-wood termites swarm through the air, seeking new colonial sites, hopefully not in your house walls. Their wings, used only during the swarming stage, break off after invasion. Look for these ½-inch-long, translucent wings in spider webs and along the rafters in your attic. If you find them, call in two or three professional exterminators for their evaluations.

Subterranean termites build mud tubes from their underground nests (they require ground moisture) to a house's wood underside. Their colonies may number a hundred thousand chewers. By all means, destroy the tubes when you find them, but then call for professional help—otherwise you may have some sleepless nights thinking about those tubes being rebuilt.

MAINTAINING YOUR SIDING

In spring and autumn, it pays to put in a few hours on general siding maintenance. By cleaning your siding and making sure it's properly sealed and caulked, you can keep it in top shape.

Cleaning

Hose down your siding to remove accumulated grime. If necessary, brush it with a carwash brush that attaches to a hose. Scrub all areas.

Touch up with paint or stain any areas that are chipped or peeling. Most wood siding is especially vulnerable to rot when the finish deteriorates. See page 99 for information on finishing.

Your particular climate may demand special vigilance. Salt-water air, for example, will corrode exposed metal quickly. Combined heat and humidity will mildew wood. If mildew is a problem in your region, retard its growth by washing wood

with a solution of ⅓ cup detergent, ⅔ cup trisodium phosphate, and 1 quart household bleach in 3 gallons of water. Brush or sponge it on well. Protect your eyes with goggles, your hands with gloves, and your plants and shrubs with a plastic tarp.

Caulking

Caulking, an important part of maintenance, excludes moisture and reduces air infiltration into your house. Caulking compound eventually dries out and needs renewal. Clean out old compound with a knife or chisel before recaulking.

Types. Latex or butyl caulking compounds are both good choices for household use, although butyl is slightly better. With its rubber-and-oil base, it fills fairly large holes, takes any type of paint, and lasts longer. It will adhere to aluminum and steel.

Caulking compound comes in three forms. Tubes that fit caulking guns are the most popular dispensers and the easiest to use for applying an even bead. Compound is also available in a can; you apply it with a putty knife. This method is typically used for filling large gaps. Rope caulk, in a spool that you unwind

and press into cracks, is handy for filling large crevices where access is difficult.

Some compounds are available in tinted shades, which might save you the trouble of touchup painting.

How and where to caulk. Practice your caulking skills at the back of the house. To use a caulking gun, cut the tube end at a 45° angle. Apply steady pressure to get a good bead, pulling the caulking gun along rather than pushing it (see drawing below). Seal the open hole in the caulking tube with a nail between uses.

Check the frames around windows and doors, especially along their tops. Lay a bead of caulking compound where the frames meet the walls (do all four sides of each frame). Then inspect within the frames of windows and doors. Caulk any cracks you see. Inspect carefully between the track of a sliding glass window or door and the sill it rests on or the jamb it hangs on.

Also examine wood-based sidings for shrinkage between boards or sheets. Where shrinkage has opened up cracks, caulk to prevent the entry of water.

Caulk wherever there are protrusions through the siding—around hose bibs, the electrical meter, pipes,

Caulking methods

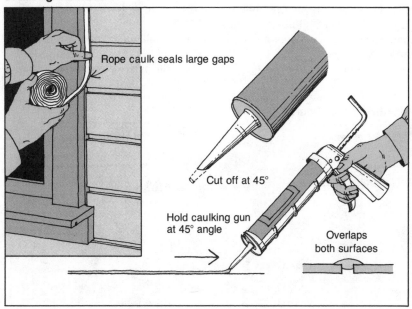

Rope caulk seals large gaps

Cut off at 45°

Hold caulking gun at 45° angle

Overlaps both surfaces

nails holding downspouts, framing members that penetrate the siding, and so forth. Also renew caulking compound where the siding adjoins a deck or masonry fireplace.

REPAIRING YOUR SIDING

Errant baseballs, careless drivers, aimless downspouts: these are just a few of the foes of siding. For the most part, such antagonists do their dirty work in a specific area. They break, dent, splinter, and crack what is—under normal circumstances—perfectly sound siding.

Most siding materials are easily repaired or replaced. Some repair jobs you can handle; others should be left to the pros. Here you'll find techniques for the repairs you can do yourself.

Repairing wood-board siding

With wood-board siding, repairs usually involve fixing warps, splits, holes, and other minor damage or replacing boards where damage is fairly extensive.

Split boards are easy to repair. You just pry apart the pieces on either side of the crack so you can liberally apply waterproof glue; then join them tightly and nail both halves to the backing (see drawing) or screw them together.

Small holes in wood-board siding can be filled with wood putty, available at lumber and paint stores. Putty comes in a variety of shades for matching lightly stained woods.

Fill the hole with putty and allow it to dry completely. If the hole is fairly large, apply the putty in layers, letting each one dry completely before applying the next. When the final coat is dry, sand the surface smooth. Then paint or stain to match the surrounding area.

Peeling paint on wood-board siding can result from a variety of causes: poor paint, improper surface preparation before painting, harsh sunlight over a long period of time, or improper wall ventilation. Except in the last case, the problem can be remedied with a proper paint job (see pages 99–102). Ventilation is another matter, one that depends directly on your climate and the presence or absence of a vapor barrier in the wall. For more on this subject, see page 103.

Buckled or warped boards usually show up where boards have been installed too tightly. If a board has nowhere to expand when it swells with moisture, it buckles. To straighten a buckled board, first try to pull it into line by driving long screws through it into the wall studs (countersink screwheads and fill). If that doesn't work, you'll have to shorten the board.

First pull out or cut off nails within the buckled area, and continue toward the nearest end of the board.

Pull the end of the board outward and shorten it slightly, using a rasp or coarse sandpaper to remove some of the wood from the end. Renail in place.

Replacing lengths of board siding. Sometimes a board is so badly decayed or damaged that no simple repair can save it. Replacement is time-consuming but rather simple.

You will need replacement boards of the same size as those being replaced, plus a square, a tape measure, a saw (preferably a back saw), a pry bar or chisel, nails, and—for some jobs—a nail puller, some small wedges, and a tube of asphalt cement.

The exact approach you should take will depend on the milled design of your siding and how it was originally nailed down. Refer to the chart on pages 110–111 to determine which pattern of siding is on your house.

For lap-type siding, first mark for saw cuts on each side of the damaged area. Cuts should be centered over wall studs. If the damage lies near a joint between two lengths of siding, you'll need to make only one cut. Use a square when marking to keep lines at right angles.

With a pry bar or stout chisel, pry up the bottom edge of the board to be removed. Drive small wedges underneath the board near the marks to keep it raised slightly.

Cut through the board along both marks. You may want to tack small blocks to the lower edge of the board above and to the face of the board below to avoid damaging them with the saw. If necessary, finish the ends of the cuts with a keyhole saw or chisel.

Break the damaged board out—in pieces, if possible, to avoid putting too much pressure on the board overlapping from above. You may find it necessary to pry up the board above the damaged one slightly to free the last bits and pieces.

Patch any cuts or tears in the building paper with asphalt cement. Then measure the width of the opening and trim the replacement

Repairing a split board

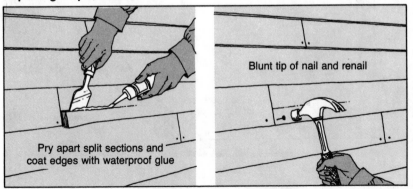

Pry apart split sections and coat edges with waterproof glue

Blunt tip of nail and renail

Replacing lap siding

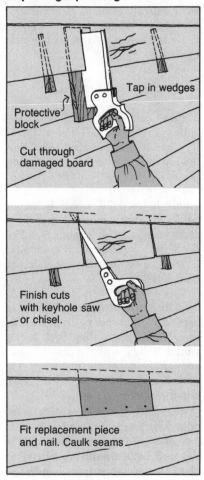

Tap in wedges

Protective block

Cut through damaged board

Finish cuts with keyhole saw or chisel.

Fit replacement piece and nail. Caulk seams

board to the right length. For greatest accuracy, measure across both the top and the bottom of the opening. Drive the replacement into exact position, hammering against a block placed along its lower edge. You may have to pull nails that pass through the board above, or cut them off with a hacksaw blade.

Nail down the replacement, using the same method as was used on the surrounding siding. Caulk the end seams and stain or paint the replacement to match the house.

Shiplap, rustic, and similar overlapping styles of siding are face-nailed to studs or sheathing. Saw cuts may be made anywhere along a length of siding with an underlayment of sheathing; where no sheathing was used, saw cuts must be centered over studs. Mark cut lines with a square to ensure their being at right angles.

If nails are in the way of your saw cuts, you'll have to remove them with a "cat's paw" or nail puller. Then make the cuts with a circular saw, setting the blade to cut not quite through the siding.

Using a chisel, pry up the damaged board at the saw cuts. When nailheads rise, pull them. Remove the board.

Repair cuts in the underlayment or building paper with a tube of asphalt cement. Trim the replacement to fit. Face-nail it in place and caulk the seams and nailheads. Paint or stain.

Tongue-and-groove siding is replaced in much the same way as shiplap, except the boards are blind-nailed and fitted together more firmly. Refer to the sequential illustrations below for replacing tongue-and-groove siding.

Pull all exposed nails located in the area to be removed, using a "cat's paw" or nail puller. Next, cut out the defective piece. It's easiest to make the necessary cuts with a circular saw; set the blade depth just shy of

the siding's thickness. Mark your cut lines, then saw just to the end of each line, dipping the moving blade down into the wood to start each cut. Be careful not to cut into adjacent boards. Also, hold the saw firmly—making this type of cut sometimes causes the saw to kick back.

Finish the cuts with a chisel. Then, using the saw, rip along the center of the length of the damaged section, again completing the cuts at both ends with a chisel.

Cave in the board and remove both halves. Before inserting the replacement piece, cut off the back edge of the grooved side so you can fit the board in place. Also repair cuts in the building paper or underlayment, using a tube of asphalt cement.

Face-nail the replacement board in place; countersink the nailheads and fill the holes with putty or caulking compound. Also caulk along the end seams. Stain or paint.

Board-and-batten siding is the easiest to replace. The battens are easy to pry off and, once removed,

Replacing tongue-and-groove siding

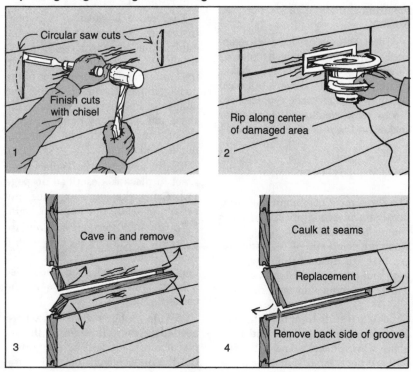

1. Circular saw cuts — Finish cuts with chisel

2. Rip along center of damaged area

3. Cave in and remove

4. Caulk at seams — Replacement — Remove back side of groove

allow easy removal of the boards. To remove battens and boards, just pry them up far enough to raise the nailheads, then pull the nails. When prying, it's sometimes a good idea to place a block beneath the pry bar to provide leverage and protect the good siding.

Repair any tears in the building paper or underlayment, using asphalt cement; then replace with identically sized boards or battens. Seal all joints with caulking compound, and stain or paint.

Replacing shingles and shakes

When a shingle or shake splits, curls, warps, or breaks, take it out and replace it. Shingles or shakes applied in "double courses" usually have exposed nails. To replace these, simply pry out the exposed nails, remove the damaged piece, install the replacement, and renail.

Shingles and shakes applied in "single courses" have their nails concealed under the shingles of the course above them. Methods for replacing these damaged shingles or shakes are the same as those used for repairing shingle or shake roofs; see page 74.

A new shingle or shake will generally turn the same aged color as surrounding shingles or shakes within half a year. If that's not soon enough, you can trade replacement pieces with weathered shingles or shakes on the rear of your house.

Exterior plywood repairs

The three most common repair jobs involving plywood are fixing areas where the surface is "checking" (showing small splits), regluing layers that are delaminating (separating), and—for more extensive problems—replacing damaged sections.

Checking results from the tendency of surface veneer to split and peel. Sand down checks and fill with putty, if necessary. Sand once more,

then refinish, using paint or stain, to match existing siding.

Delamination—where the wood plies separate—is a problem to watch for each spring and autumn. To repair edges just beginning to delaminate, apply waterproof glue between plies and then nail them down. For large-scale delamination, replace the entire sheet (see below).

By the way, when installing plywood siding, you can try to anticipate the problems that cause delamination by keeping plywood dry before applying it, waterproofing the edges, caulking (page 93) or flashing the joints (page 106), and finishing the sheets adequately with stain or paint.

Damaged plywood. Don't try to repair plywood that has been broken or badly damaged; replace it. You can replace either the entire sheet or part of it. Most often, it's easier to replace the entire sheet; otherwise you'll have to provide 2 by 4 backing around the edges of the replacement (unless the siding is already backed by sheathing). To replace damaged lap panels, use the method described for replacing wood-board lap siding on page 94.

To remove a sheet of plywood, first remove any battens covering the joints or nailed as trim over the plywood. Wedge a sturdy chisel into the joints and pry up the edges of the damaged panel enough to lift the nailheads. Pull the nails. Remove the sheet and repair any damage in the underlayment or building paper, using asphalt cement.

Mount and nail the replacement sheet in place as described on page 112. Caulk the joints, replace any battens or trim, and paint or stain.

To cut out a damaged section, use a circular saw. Mark all cuts with a square, and plan vertical cuts so they'll be centered over studs. Use a back saw to finish cuts at the corners.

Provide 2 by 4 backing where needed so that all edges of the replacement piece will be supported. Cut the replacement piece to fit the

cutout. Repair the building paper with asphalt cement, if necessary. Nail the replacement piece in place, caulk all seams, and paint or stain.

Repairing hardboard

Typical problems with hardboard siding include small holes, buckling, stains, and damaged areas. Following are methods for repairing these.

Small holes. Fill small holes in hardboard with wood putty, then sand and paint. If a hole is fairly deep, build up the patch gradually, filling two or three times before the final sanding.

Buckling in hardboard siding is caused by improper nailing or moisture. Check the nailing first. Is the siding nailed with box nails long enough to penetrate the studs 1½ inches, and properly spaced? If not, renail properly.

If outside moisture hasn't been repelled by paints, it may be the culprit. If this is the case, paint. Does the wall have a moisture barrier to check water vapor from inside the wall? See more about vapor barriers on page 103.

Stains on hardboard can be removed with a mild detergent or—for oil-based stains—a solvent.

Major damage requires removal of a panel. Follow the instructions given for plywood (preceding).

Repairing vinyl siding

Any section of vinyl siding that is cracked or punctured should be replaced. To do this, you'll need a special vinyl siding tool called a "zipper." The best time to do the work is during warm weather, when the vinyl is pliable. (Vinyl becomes brittle and prone to crack in cold weather.)

If your siding is vertical, lift the panel adjacent to the damaged piece.

If you have horizontal siding, lift up the panel above the damaged section. This should expose the nails holding down the damaged panel. Pry out the nails. Mark cut lines on each side of the damaged area, using a square. With tin snips or a backsaw, cut the panel along these lines.

Remove the damaged piece and install a new piece that's 2 inches longer—1 inch for overlapping each adjacent panel—or only 1 inch longer if it ends at a corner or joint. Snap the new panel into place, overlapping the cut ends of existing siding. Nail it in place with the same type of nails you pulled out of the damaged piece. Then use the zipper tool to snap the locking bottom of the above or adjacent panel back into position.

Repairing aluminum and steel sidings

While prone to dent, metal sidings are resistant to most other kinds of damage. Because aluminum is considerably softer than steel siding, it is damaged more easily—and repaired more easily. If your steel siding is so badly damaged that sections need replacement, we suggest you call a professional. Aluminum is different. Because you can cut it fairly readily, you can make your own repairs.

Following are techniques for repairing dents, scratches, and corrosion in metal sidings, and methods for replacing sections of aluminum.

Small dents can be pulled out by drilling a hole in the dent, screwing in a self-tapping screw with two washers under the head, then gently pulling on the screwhead with a pair of pliers. Fill the hole with plastic aluminum or steel filler, available at paint and hardware stores. Follow directions on the tube. Sand smooth and touch up with matching paint.

Scratches and corrosion. Scratched aluminum can be touched up with a metal primer and latex house paint to match the manufacturer's finish. Repair corroding steel by cleaning

the rust off with steel wool. Prime the area with rust-resistant metal primer before applying a topcoat of latex house paint.

Damaged corner posts of aluminum siding can be replaced. Score along the outer edge of the groove (see drawings below). Then bend the aluminum with pliers to break. Cut a new corner post as shown, along the inner groove, and attach the new post with two rivets at the top and two at the bottom.

For major damage to aluminum siding, use a utility knife to slit the damaged panel just above center as shown. Cut the length of the damaged area, then straight down at both ends of it. The section will drop when you have made the cuts.

Then cut a piece of new siding 6 inches longer than the piece you removed, so that it will overlap the existing siding by 3 inches on each side. Cut off the nailing strip so it will tuck up under the butt of the course above. Liberally coat the area

Replacing aluminum corner posts

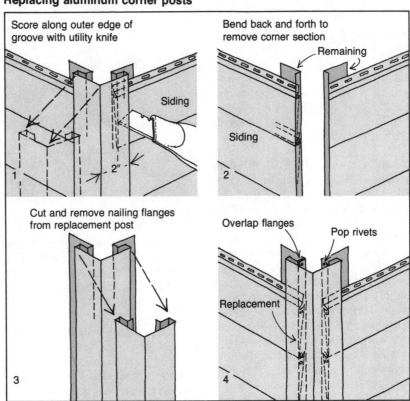

1. Score along outer edge of groove with utility knife — Siding — 2"

2. Bend back and forth to remove corner section — Remaining — Siding

3. Cut and remove nailing flanges from replacement post

4. Overlap flanges — Pop rivets — Replacement

Replacing damaged aluminum panels

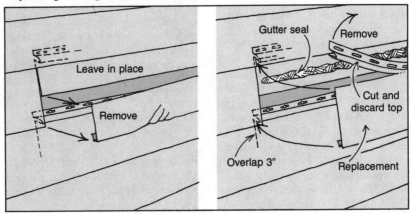

Leave in place — Remove

Gutter seal — Remove — Cut and discard top — Overlap 3" — Replacement

above the opening with butyl gutter and lap seal.

Lock the new piece in at the bottom and press it into the gutter seal. Hold the piece firmly until the seal has hardened well. If necessary, wedge a 2 by 4 between the patch and the ground and let it set overnight.

Repairing stucco

If cracks or holes appear in stucco, the original stucco mixture may have been unsatisfactory, the building may have settled, or your house may have lived through an earthquake or two.

Hairline cracks can be covered with a coat of latex paint. Larger cracks should first be filled with a latex caulking compound, then covered with latex paint.

Small holes—up to a width of 6 inches—are easily repaired. First scrape out the debris, then staple in some fresh wire mesh if the existing mesh is damaged. Mix up a small batch of stucco patch (available at lumberyards and hardware stores) and apply it. Before applying the patch, wet the area of the hole thoroughly so that the old stucco and wood lath, plywood, or building paper won't weaken the new stucco by sucking out its moisture. When mixing the stucco patch, follow manufacturer's directions. Use only enough water to wet the materials thoroughly. Mix the water into the materials vigorously.

Holes larger than 6 inches across should be repaired by the same methods as those used in applying new stucco. First, prepare the area as described above for small holes. Then apply an initial coat of stucco to within ½ inch of the surface; the stucco should have a consistency that allows it to set up stiffly on the wire mesh. While this first coat—called a "scratch coat"—is drying, scratch it vertically and horizontally with a nail. The scratches will allow the second coat to adhere tightly to it.

Repairing stucco

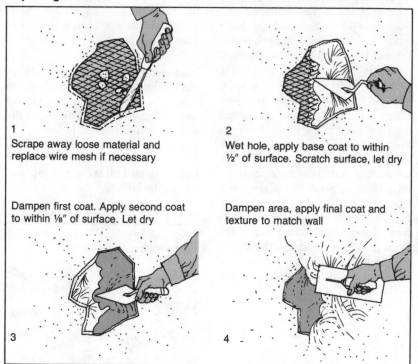

1. Scrape away loose material and replace wire mesh if necessary

2. Wet hole, apply base coat to within ½" of surface. Scratch surface, let dry

3. Dampen first coat. Apply second coat to within ⅛" of surface. Let dry

4. Dampen area, apply final coat and texture to match wall

After the scratch coat cures for a day, apply the "brown coat," so named because the sand and cement used are brown instead of white or pigmented as in the final coat. Dampen the surface and apply the brown coat to within ⅛ inch of the surface. Allow the brown coat to set up for a day.

Finally, dampen the area and apply the final coat (made of white Portland cement and white sand mixed in the same 1-to-3 proportions as the earlier coats). Bring this coat flush with the surface of the wall and match its texture to the surrounding stucco when possible. If necessary, use a scraping plank or trowel to clear excess stucco from the surrounding wall. The final coat may be either pigmented or painted to match the wall.

Repairing asbestos-cement shingles

Before you attempt to replace shingles on an asbestos-cement shingle wall, check to see if new shingles are available in your area.

Measure the shingles that need replacement and have your supplier cut the replacements to fit. (Dealers have a special cutting tool for this purpose.)

To replace a shingle, pull the nails holding it in place and remove it. Slide in the new shingle and face-nail through the designated nail holes. Drive each nail to within ½ inch of the surface, then use a nailset to pound it flush. Use corrosion-resistant box nails long enough to penetrate 1 inch into sheathing or studs (see drawing on next page).

If there are no nail holes, predrill them yourself, 1 inch above the top of the underlying shingle and 1 inch from each edge of the replacement.

Wear a face mask. Asbestos shingle fibers are dangerous to breathe. It's also a good idea to wear eye protection; the shingles are very brittle.

If many shingles need replacement, consider knocking them all off and replacing them with new siding. You might even find that the siding underneath is salvageable. Remove asbestos shingles by striking and breaking them with a ham-

Replacing asbestos-cement shingles

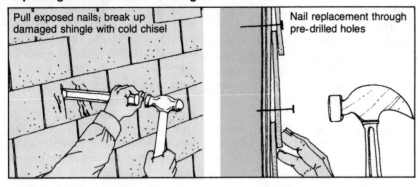

Pull exposed nails; break up damaged shingle with cold chisel

Nail replacement through pre-drilled holes

mer or pry bar. Pry loose the pieces and let them fall to the ground.

PAINTING AND STAINING

For many sidings, a proper paint job or other finish is the first line of defense against the elements. If its exterior paint is cracked and peeling, a house not only looks shabby but is also vulnerable to the assaults of severe weather.

Recommendations for specific sidings

Each particular siding material has its own finishing requirements—or the lack of them. Producers of wood siding products supply instructions or recommendations for finishes. Synthetic sidings are usually factory-finished. Here are some of the basics.

Wood boards—except for redwood, red cedar, and Southern red cypress—need to be stained or painted.

The three exceptions are so naturally resistant to water rot and insect damage that nothing need be done to preserve them. But keep in mind that, untreated, they will acquire a natural patina with age. You can minimize the change by treating them with a clear waterproofing sealer, though this may tend to darken the natural color slightly. Or you can accelerate the natural aging process with wood bleaches, available at paint supply stores. These

permanently change the wood's appearance to a weathered gray.

To be sure you like the results, test any finish on a piece of wood of the same species before applying it to your house.

Any type of wood left untreated can be marred by unexpected stains. Typical stains result from drainage off roofs, spattering mud from the ground, and sap falling from trees. Using corrosive nails in siding can also cause staining.

Woods not naturally resistant to decay require a protective stain or paint. Clear, light-bodied wood stains tend to preserve the grain and texture of wood. More heavy-bodied stains produce a more uniform color, preserving the texture but obscuring the grain of the wood.

It's a good idea to brush some type of clear waterproofing sealer or protective finish on the ends of *all* wood boards. This is where water penetrates a board most easily. And when painting wood-board siding, apply an initial prime coat and then a finish coat on all new woods.

Shingles and shakes usually give decades of service without any attention. The wood may acquire a dark gray, silver, or dark brown patina—depending on climate and moisture, which also affect the speed of the weathering process.

Simulating the aging process by turning shingles and shakes a weathered gray, bleaches allow you more control of coloring. When using wood bleach, follow the manufacturer's suggestions for brushing

it on and be careful to protect your eyes, skin, and landscape plants from the bleach. Test bleach on a shingle before applying it to your walls.

Wood preservatives with a fungicide and mildew retardant, available at paint stores, are sometimes desirable in climates where heat and humidity cause wood materials to deteriorate quickly.

Both transparent and opaque stains are sometimes used on shingles and shakes. Once you apply pigmented stain, though, plan to re-treat your siding every few years. Use latex-base stains.

Paint is sometimes a problem on shingles and shakes; moisture absorbed through their sides and backs can cause the paint to peel. And water-soluble extracts in cedar can bleed through paint. So if you're using a latex paint, a special stain-blocking latex primer coat should be applied first. Some shingles come factory-primed.

Plywood. Top grades of textured plywood can be finished with a light-bodied, semitransparent oil base stain to preserve both the grain and the texture of the wood.

Most homeowners stain other grades of textured plywood with a heavier-bodied, latex base opaque stain that obscures the differences in color between the veneer and repair patches. Opaque stains hide the grain but retain the texture of the wood.

Paint can also be used over textured plywood and is the only finish recommended for sanded plywood or plywood overlaid with resin.

Because some woods, especially redwood, have natural stains that can bleed through latex paint, first apply an oil base primer or a specially formulated stain-blocking acrylic latex primer. For the top coat use a good quality acrylic latex paint.

Redwood and cedar plywood, like board siding, can also be bleached to a driftwood gray.

Hardboard. Most sheets and lap panels of hardboard are made to be finished with acrylic latex paints. Usually, manufacturers recommend an oil or acrylic latex prime coat and then one or two top coats of acrylic latex paint.

You can buy hardboard unfinished, primed, or factory-finished with both a prime coat and top coats.

Some hardboards are factory-finished with an overlay of fibrous phenol resin, vinyl, or acrylic. Like regular hardboard, these overlaid hardboards are usually finished with acrylic latex paints.

Manufacturers of hardboard sometimes sell nails with colored heads, touchup paint, and color-matched caulking compound for a complete finishing job.

Vinyl siding requires no finish. Its color goes right through the material, so scratches don't show. Maintain vinyl panels simply by hosing and sponging with a mild liquid detergent every few months.

If you want to change the color of your existing vinyl siding, you may have a problem. The ability of vinyl to receive and hold paint satisfactorily is debatable. Check with the manufacturer of your vinyl siding for recommendations on painting.

Aluminum and steel. Factory-finished aluminum and steel sidings have a coating superior to anything that a homeowner can apply. Usually, the factory finish—a baked-on enamel or an acrylic or vinyl overlay—is guaranteed for five years.

When repairing aging metal siding, clean the surface thoroughly, spot prime any areas where metal is exposed with a good quality metal primer, and follow with a coat of latex house paint.

Painting know-how

Painting is a task almost all homeowners can accomplish, especially those who live in a one or two-story house. In houses exceeding two stories or situated on a steep hill, it may be best to have the job done by a professional. But in the vast majority of cases, if you take your time, you can do the job as well or better—and save money.

The evolution of painting into a fairly easy do-it-yourself skill began with the paint roller and easy-to-apply paints, and has led to the paint sprayer. For most exterior paint jobs, here are some helpful tips.

Preparation. On exterior wood surfaces, it is essential to remove old paint if it is peeling, blistering, or flaking. Sanding (best on smooth wood surfaces) and scraping (easy but time consuming) are the techniques best suited to the home craftsperson. You can also apply a chemical paint remover, but it's expensive. Burning with a blowtorch is a technique that absolutely should be left in the hands of an experienced professional. Many a bold homeowner has accidentally blowtorched his house to the ground. Elbow grease is safer.

After scraping or stripping your house, wash it with a mild detergent, then hose it off. Allow the house to dry thoroughly before painting. Caulk with a paintable caulking compound (page 93), and renew any aging window putty before starting to paint.

On metal surfaces (aluminum or steel), use sandpaper or a stiff wire brush to remove loose paint and corrosion. (On steel, a chemical rust retardant should be applied either before painting or in combination with the paint.)

On masonry sidings use a stiff wire brush for loose, peeling, or badly chalked paint. An alternative method (although expensive, dirty, and likely to upset neighbors) is sandblasting. You can rent the equipment or hire professionals. If you have a painted masonry surface that is only slightly chalked, it can be covered with a sealer and then repainted.

Both in preparing the siding and in painting it, use plastic or cotton drop cloths to protect fences, decks, patios, and shrubs adjacent to the house.

Applying the paint. Once you've chosen the proper paint or finish for your siding (see the chart on page 102), it's time to gather the necessary tools and begin. You can apply paint by brush, roller, pad, or spray. Helpful painting tools are shown in the drawing on the facing page.

Be sure to get a high quality 2-inch trim brush and a 4-inch brush for painting wider surfaces. If your siding material has large, flat areas, you can paint with a 9-inch roller. Get a roller with a fine nap for smooth surfaces; a thicker nap is better for heavily textured surfaces. Buy an 8-foot extension for the roller so you can reach high areas. When using a roller, apply heavy pressure to work the paint well into the surface.

Or you can go the route of the paint sprayer. For more about this option, see ''Rental tools,'' below.

You'll also need a ladder or some scaffolding. You can rent scaffolding equipment, as discussed below.

Exterior painting is best done during fair, dry weather with moderate temperatures between 50° and 90°. Wait until the morning dew has evaporated, and stop painting before evening dampness sets in. Avoid painting during windy or dusty weather, particularly with slow-drying solvent base paints. (If insects get caught in wet paint, brush them off after it dries.)

Remove shutters and screens and paint them on a flat surface first. Paint windows, trim, and doors next. Paint the walls last so you won't have to lean a ladder against their newly painted surfaces to do the windows and trim. Start at the top and work down, painting in the direction of the grain if you're painting wood.

Rental tools. A thriving industry rents all sorts of specialized equipment for home craftspeople (and contractors as well), and many time-saving tools are available for house painting.

Paint removers. Several rental devices relieve the tedium of preparing the outside of the house for painting.

The simplest mechanical aid for removing paint in small quantities

Painting equipment

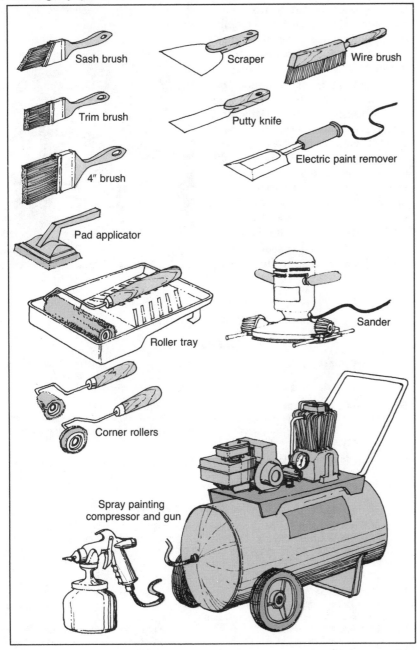

Sash brush

Scraper

Wire brush

Trim brush

Putty knife

4" brush

Electric paint remover

Pad applicator

Roller tray

Sander

Corner rollers

Spray painting compressor and gun

and-batten siding. But like sand-blasting, it presents almost as many problems as solutions. On textured plywood, brush rather than spray on the prime coat of paint or stain to ensure that it works its way well into the wood fibers.

Outdoors, the equipment can be used only on windless days because of the tendency of the aerated paint to drift about the neighborhood and alight on cars, windows, and buildings, sometimes hundreds of yards distant. Further, the equipment is difficult to keep clean, particularly if quick-setting water-base paints are being used. The paint tends to clog the spray head and build up in the hose. Some rental agencies have despaired over the clogged equipment they get back from renters and no longer rent sprayers for other than mineral-solvent paints. Rental rates vary according to size. Be sure to buy or rent a respirator and goggles.

A promising replacement for the pneumatic spray gun is the airless gun. With this device, air pressure directed into a holding tank forces paint into a single hose leading to the spray nozzle. When the operator presses a button, the paint is forced out under pressure, atomized by the nozzle into a wide cone. Since it is not mixed with air, the paint drifts very little.

The gun is easy to keep clean. At the end of the day, the operator simply leaves the spray head, still attached to the hose, immersed in paint thinner or water. When the operator starts up in the morning, he needn't worry about a clogged nozzle or choked hose—the paint in the hose remains as fresh as in a can.

The airless spray gun, which is rapidly displacing pneumatic sprays in the rental market, does have a few drawbacks: it requires grounding, it is lavish in its output of paint, and it can be very dangerous if handled improperly. At close range, the nozzle pressure is high enough to pierce a painter's skin.

Scaffolding. The timesaving and safety factors that make scaffolding useful in heavy construction can also apply to jobs around the house, par-

is an electric paint remover that applies heat to a small area—about half a square foot—causing the paint to shrivel and fall off.

A smaller version of the same tool melts putty, eliminating the need to scrape it off each window frame and risk breaking the glass.

Next in efficiency is a power sander that scuffs off swaths of paint with an abrasive disc. The machine is adjustable, so you can set the exact

depth to which you wish to sand. This is an effective tool for removing old paint. The renter purchases a copious supply of the abrasive discs, which wear out rapidly, at the time he rents the tool. Any that are not used can be returned for credit when the tool is checked in.

Spray painting is often the only way to cover a rough-textured surface such as shingles, fencing, stucco, acoustic plaster, or board-

ticularly if you're starting a complete repainting or re-siding project.

You can rent scaffolding for a relatively low fee, considering the time you'll save by not having to climb up and down a ladder every few minutes to reposition yourself. It allows you to stand on a level platform with both hands free. There's plenty of room on each level for paint and tools, and you can work on larger areas of the house without stretching or leaning dangerously. Guardrails, cross braces, and end frames also contribute to safety.

Aluminum scaffolding is sometimes available as an alternative to steel. Lightweight and easy to move, it simply unfolds into place. It is more expensive, however. In general, for a single family residence, scaffolding runs about $100 a month, not including planking; but prices vary and depend, of course, on the amount of scaffolding required.

Once at the site, scaffolding simply bolts or clamps together. Be sure to get assembly instructions and safety precautions from the rental agent. If your lot is especially steep or if you want to put up a scaffold over stairways or other obstacles, you should hire an experienced worker to set it up for you; the potential risk isn't worth the modest savings.

For a very simple scaffold system, you can get by with two ladders, two ladder jacks, and a scaffold-grade plank to lay between them. Such a system is easily maneuvered but has no safety rail.

Paints and Stains

Interior-Exterior Paints

A. **Latex masonry**–Dull finish. Water thinned. Very good durability. Quick drying. Easy application and cleanup. Requires a special primer when used over old, chalky surfaces. May not adhere well over several old coats of oil/alkyd paint.

B. **Powder cement base**–Powder to be mixed with water. Least expensive masonry paint. Can be applied to any porous masonry surface—such as concrete block, nonglazed brick, stucco, or cinder block—which is either unpainted or previously painted with the same kind of paint.

C. **Acrylic enamel**–Glossy, hard enamel finish. Stain and chip resistant. Bare surfaces should be primed before application. Available in quart cans or spray containers.

Exterior Paints and Stains

D. **Oil house paint**–Oil or alkyd base. Solvent thinned. Gloss finish. Often thicker than latex base. May require fewer coats than latex base in repainting. Slow drying. Bright colors. Also available as combination house and trim paint. Use is restricted in some regions because of pollution from evaporating solvents as paint dries.

E. **Latex house paint**–May be modified acrylic or vinyl emulsion water base. Dull or semigloss finish. Highly resistant to alkalis, blistering, and sun fading. Easier application and cleanup than oil base. No fire hazard. Can be applied in humid weather and to damp surfaces. Less hiding power than oil base; may be more expensive than oil base in the long run. Also available as combination house and trim paint.

F. **Exterior trim**–High gloss. Hard, durable surface. Available in oil and alkyd base as well as latex base. Latex base varieties can also be used on brick or asbestos shingles. Excellent color retention and durability because of hard surface. Latex trim quicker drying than oil-base trim.

G. **Shingle**–Available in oil or alkyd resin base as well as latex base. Coating breathes and so should not blister or peel. Latex variety may need special primer.

H. **Deck/Porch enamel**–High gloss. Oil and alkyd base. Solvent thinned. Excellent adhesion to floor. May be susceptible to deterioration on concrete because of alkalis or excessive moisture.

I. **Exterior wood stains**–Natural wood finish available in semitransparent (subtle color) or opaque (hides grain pattern). Special shingle stain available. Oil, alkyd, or latex base. Penetrates and stains wood; won't peel, blister, or chalk like paint. Use with sealer on new surface or previously stained surface.

Surface	Appropriate Paint/Stain
Walls	
• Wood siding	D,E,G,I
• Stucco, brick	A,B,E,F
• Concrete block	A,B,E
• Wood shingles	G,E,I
• Asbestos shingles	E,F,G
Wood Trim	
• Window frames	C,D,E,F,I
• Doors, door frames	D,E,F,I
• Eaves	D,E,F,I
• Decks, porches	D,H,I
Metal Trim	C,D,F

Note

Some enamels, metal primers, and clear finishes, available in either brush-on or spray form, contain rust inhibitors.

PREPARING WALLS & INSTALLING SIDING

When your siding is in such bad shape that it demands replacement, or if you're doing some remodeling that calls for a new exterior wall covering, nothing less than a complete siding job will do.

But before you jump in with both feet, look through the following pages to gauge how much work will be involved and what tools and skills it will require. If it looks as though you may be getting in over your head, refer to page 90 for information on working with a pro.

Consider the material. Plywood and hardboard can be installed fairly swiftly. Wood boards require more skill and time. Shingles and shakes are tedious but fairly easy to put on.

The three popular synthetic sidings—aluminum, steel, and vinyl—are among the most difficult to apply because of the considerable cutting and fitting necessary. Of the three, vinyl is the easiest, aluminum is a bit more difficult, and steel is best left to a professional familiar with its installation.

Patching stucco, as described on page 98, is well within the grasp of the handy homeowner. But applying stucco to an entire house siding should definitely be contracted to a professional. It's a skill that requires years of practice.

INSULATION, VAPOR BARRIERS, AND VENTILATION

Heat always moves to a colder location. In winter, the heat in your house passes through walls, roofs, and floors—a movement that causes problems. For one, your utility bill rises as the heat escapes. For another, as the heat goes through, the moisture-laden room air condenses on the cold inner faces of exterior surfaces, where it accumulates. Eventually, it blisters the outside paint, forms stains inside, saturates insulation, and damages the house's structure.

To retard the flow of heated air and moisture, and to minimize their effect, walls should be properly insulated, equipped with vapor barriers, and vented.

If you are opening up walls for a re-siding project, it may be the perfect time to take care of these important tasks.

Insulation

If the siding has been removed, exterior walls are easy to insulate with insulation batts or blankets. And because batts and blankets come with a vapor barrier attached to one face, you can kill two birds with one stone by using them.

If you don't plan to open up walls, you can hire a professional to blow in or foam in insulation through scores of holes drilled through siding. Or you can fasten 4 by 8-foot rigid insulation boards over the old siding (or over studs) and then apply your siding over these sheets. Another alternative is to put on synthetic siding with slip-in insulation panels.

For a complete look at insulating, see the *Sunset* book *Do-it-yourself Insulation & Weatherstripping*. Also refer to the information in the roofing section, on page 62.

Vapor barriers

To keep moisture from passing through insulation and collecting inside walls, install vapor barriers. They repel moist air before it gets to the cold part of a wall, where it would condense.

When installing a vapor barrier, remember this very important rule: *Put the vapor barrier toward the warm-in-winter side of insulation.* Unless you are in an extremely warm and humid climate, this means that you should face the vapor barrier in toward the room, on the back side of interior walls.

If you can't install insulation batts or blankets that have attached vapor barriers, put in a separate barrier or rely on ventilation to handle the problem, as discussed below.

A good separate barrier consists of either 2-mil (or thicker) polyethylene, laminated asphalt-covered building paper, or foil-backed gypsum wallboard on interior walls.

You can also try painting interior surfaces with two coats of a paint that has a "perm rating" of less than one. If you can't find such a paint, use two coats of high-gloss enamel or a varnish-based pigmented wall sealer, followed by a coat of alkyd paint.

Another alternative is to hire a contractor to foam in insulation. Foam insulations create their own vapor barriers.

Ventilation

If a house is properly ventilated, a vapor barrier may not be needed (particularly in a dry climate). The warm, moist interior air takes the easiest route to the cold outside—in this case, through vents or fans.

In addition to the vents discussed on page 63, you can buy small siding vents, designed to be pushed into small holes drilled in siding or wedged between siding boards. These ventilate inner wall spaces not protected by vapor barriers—particularly where new siding is applied over old. They keep moisture from condensing on the back side of new siding.

When re-siding, don't cover up any existing crawl space vents or access holes. Replace any damaged vent screens.

PREPARING THE WALL

The first step in installing any type of siding is wall preparation. If you're working with new construction, where wall studs or sheathing are exposed, you're ready to begin. But if the wall has a previous layer of siding on it, you probably have some preparatory work to do.

Providing a nailable base

Regardless of the type of do-it-yourself siding you're putting on, it must have a nailable base. If the wall is flat and sound, this won't require extra work. But if the existing wall is made of masonry or is bumpy and irregular, you may need to strip off the existing siding or provide a nailing base of furring strips. If you're working with new construction, you may need to put sheathing over the studs (in some situations, sheathing is also used in conjunction with furring).

Removing existing siding. If you can avoid this job, do. It's generally quite a bit of messy work, and it exposes your house to the weather. On the other hand, it's the only way to go if you want to install insulation batts or blankets inside walls; if your existing siding is aluminum, vinyl, steel, or asbestos cement; or if your existing siding is in such bad shape that it requires removal.

Equip yourself with a claw hammer, a thin-bladed pry bar, and a chisel. A pair of vise grip pliers is handy for pulling nails in hard-to-get-at places. A nail puller—a special prying device—is very handy for pulling nails quickly, but it mars wood. If you're removing wood siding that you intend to save and reuse, gently lift the siding away from the studs or sheathing with a thin pry bar, then pull the nails, protecting the surface with a scrap block.

Wood boards. The easiest wood-board siding to strip is board-and-batten. Simply pry up each batten to raise the nailheads. Pull the nails and remove the batten, then the boards.

For lap board siding (and lap plywood or hardboard), start at the top of the wall and pry off the top molding. Then you'll have easy access to the nails holding down each successive layer of boards as you work down to the bottom. Remove the nails with a nail puller or pry bar.

For wood boards with mitered edges that are face-nailed, use a nail puller to lift the nails until you can remove them with a claw hammer. If you want to avoid marring the wood, use the nail puller on the first board, and then lift subsequent boards with the pry bar to raise the nails.

Tongue-and-groove boards that are blind-nailed through the tongue can be removed by first taking off the molding and pulling any visible nails from the top board on a horizontal board wall or the end board on a vertical board wall. Then work your way down or back, using a nail puller or pry bar to raise nails from each tongue.

Shingles and shakes. If you intend to save and reuse shingles and shakes, use the nail puller carefully on exposed nails. Where nails are concealed, work from the top row down, removing one horizontal row at a time. This is usually much more work than it's worth in salvaged materials.

If the shingles or shakes will be discarded or burned for firewood, you can use a square-bottom shovel to pry them from the wall. Simply insert the shovel underneath the shingles or shakes, lift them up, and pull them off. Start at waist level and work up and down the wall.

Plywood and hardboard. With a pry bar or a nail puller, you can lift the nails holding plywood or hardboard sheets at the perimeter and along studs. A pry bar wedged underneath the sheets will not mar their surface. A nail puller will make the work go quickly, but will mar the sheets.

Vinyl, aluminum, and steel. For horizontal panels, start at the top by removing the molding that covers the last row of nails. Using the pry bar or nail puller, lift the panel and loosen the nails. Then remove the nails from each panel, slip out the interlocking device at its bottom edge, and expose the nails for the next panel. Work your way to the ground.

For vertical panels, start where the nails of an end panel are concealed by molding. Remove the molding, pull the nails, and unlock the panel. Work sideways along the wall.

Because aluminum panels bend easily, it's hard to preserve them for reuse. If you're not reusing the aluminum, it can often be recycled.

Stucco removal is hard work. For an extensive job, you might do well to call in a professional. To do it yourself, you'll need a cold chisel and a heavyweight hammer. With hammer and chisel, chop the wall off. Or consider renting an electric or air-powered hammer. Be sure to protect your eyes by wearing goggles when you work.

Asbestos-cement shingles are easy to remove, but be sure to protect your eyes with goggles, your hands with gloves, and your lungs with a face mask.

Strike the brittle shingles with a hammer—they'll simply fall apart. A pry bar can be useful for removing pieces still hanging on nails.

Furring is generally a gridwork of 1 by 3 boards or strips, placed so as

Layouts for furring walls

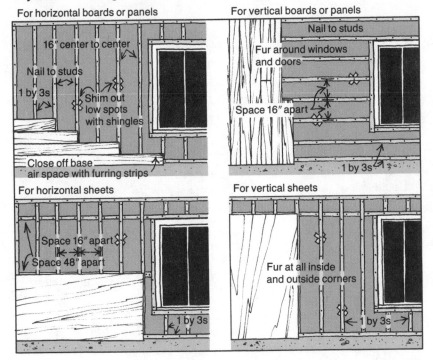

For horizontal boards or panels
- 16" center to center
- Nail to studs
- 1 by 3s
- Shim out low spots with shingles
- Close off base air space with furring strips

For vertical boards or panels
- Nail to studs
- Fur around windows and doors
- Space 16" apart
- 1 by 3s

For horizontal sheets
- Space 16" apart
- Space 48" apart
- 1 by 3s

For vertical sheets
- Fur at all inside and outside corners
- 1 by 3s

to provide nailing support for siding at the necessary intervals. Shims are placed between the furring and the existing wall covering where required to make a flat surface.

Placement of furring boards depends on the new siding. Typical layouts are shown in the drawings above.

Use a long, straight board or a taut line to determine where it is necessary to shim furring in order to provide a uniform plane for nailing. Use short sections of wood shingles for shims. Also use a level or plumb bob to make sure that furring is plumb.

Nail through the furring every 12 inches with nails that penetrate studs at least 1 inch. (Vertical furring boards should be placed directly over wall studs.)

If the existing walls are masonry, use concrete nails or masonry anchors to fasten furring boards in place.

Sheathing is used under some siding materials to give them sufficient rigidity, to provide added strength, to serve as a solid base for nailing, and to boost the insulation of walls. Check local codes to determine whether or not sheathing is required for the type of siding you intend to install.

Sheathing is generally applied only to new structures. The previous siding of older walls usually acts as sheathing for new coverings.

Several types of sheathing are commonly used: plywood (the most common), exterior fiberboard, exterior gypsum board, and ordinary board lumber. Plywood is most frequently used because its large panels are easy to apply, and usually afford enough lateral strength to eliminate the need for bracing during the framing of a house.

The chart on the next page compares the main types of sheathing and basic application techniques. All types should be nailed directly into studs. As a rule, waterproof nails should be used; they should penetrate studs at least 1 inch.

Building paper is a wind and water-resistant material, usually felt or kraft paper impregnated with asphalt, applied between siding and sheathing or studs. Constructed as a breathable membrane, building paper is a kind of insurance—another line of defense against the elements. Without building paper, a crack in siding can let in moisture that will rot sheathing or studs.

Local building codes will tell you whether you are required to use building paper, but you may want to use it even if not required. Your climate and type of siding will determine whether or not it's needed. How much wind and wind-driven rain or snow will your siding be subjected to? Does your siding consist of big sheets of relatively water-resistant material such as plywood or hardboard, or of small shingles or shakes that present many places for wind and water to penetrate?

Building paper is purchased in rolls 36 to 40 inches wide and long enough to give 200 to 500 square feet of coverage (allowing for overlap).

Apply building paper in horizontal strips, starting at the bottom of each wall and working up. Overlap 2 inches at horizontal and 6 inches at vertical joints.

Applying building paper

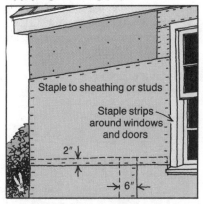

Staple to sheathing or studs

Staple strips around windows and doors

2"

6"

When applying building paper, cut it with a utility knife. Staple or nail it to studs or sheathing, using enough fasteners to hold it temporarily in place—about one fastener per foot along the top edge of each strip. Later, the siding nails will hold it permanently in place.

Preparing windows and doors

New siding will add to the thickness of your walls (unless you strip the

Wall Sheathings: How They Rate

Qualities	Types			
	Exterior Plywood	**Wood Boards**	**Exterior Fiberboard**	**Exterior Gypsum board**
Direction of application	Vertical or horizontal	Diagonal or horizontal	Horizontal	Vertical or horizontal
Panel sizes and types	$5/16$, $3/8$, $1/2$-inch thicknesses in panels of 4 by 8, 9, or 10 feet. Square-edge or tongue-and-groove.	1 by 6: end-matched tongue-and-groove. 1 by 8 or 1 by 12: shiplap.	$1/2$, $25/32$-inch widths in 2 by 8-foot panels. Tongue-and-groove or shiplap.	$1/2$-inch widths, 2 by 8-foot panels. Tongue-and-groove.
Rigidity	Good	Good	Fair	Good
Insulative value	Low	Fair	Good	Low
Nailing	Nail every 6 inches along panel's edge and every 12 inches into center supports.	Three nails per bearing for widths of 8 inches or more, 2 per bearing for lesser widths.	Use roofing nails 3 inches apart along edges, 6 inches apart intermediately.	Gypsum nails every 4 inches around edges and every 8 inches intermediately.
Does wall need diagonal bracing?	No	Only for horizontal application	Yes, with standard types	In some areas
General notes	Use Standard grade. Apply with panel ends spaced $1/16$ inch apart and edges $1/8$ inch apart.	Also available are 2 by 8-foot panels made from edge-glued lumber, overlaid with building paper.	Easy to handle and apply. Don't nail within $5/8$ inch of edges. Only a special type will serve as sole nailing base for siding.	Not a nailing base for siding.

walls first). For weathertightness and appearance, it's often necessary to build up the jambs and sills of windows and doors to compensate for the added thickness.

Extending a jamb

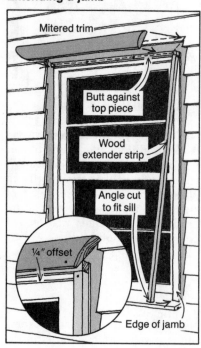

Mitered trim

Butt against top piece

Wood extender strip

Angle cut to fit sill

$1/4$″ offset

Edge of jamb

For synthetic sidings, special add-on trim pieces are often provided to take care of this job.

For wood jambs, it's a simple matter of adding small wood strips to

Extending a sill

Plane front edge of sill flat

Build out sill

Caulk

Nail

Drip groove

extend them. Gently pry off the old moldings and set them aside. Cut extender strips the same width as the jambs and a depth equal to the additional thickness of new siding (including sheathing, furring, and all).

For each window and door, cut the first extender to fit across the top, and glue and nail it in place. Then butt side jamb extenders up against it and cut their lower ends at the angle needed to match the sill.

Flashing

Flashing—specially formed galvanized sheet metal or aluminum—prevents the water that runs down a wall from penetrating the horizontal joints between materials. It's available at building supply centers and some sheet metal shops.

Before applying any type of siding, be sure that drip caps extend over windows and doors. These L-shaped metal flashings extend from under the siding out over the window and door frames. Water

Flashing for siding

Z-flashing between plywood or hardboard sheets

Drip caps over windows and doors

Dormer

2" clearance

Step flashing

Flashing

Roof

rolling down the walls is carried over the frames and drips to the ground. Cut these flashings with tin snips and nail them in place over windows and doors with galvanized roofing nails.

Along horizontal joints between large sheets of plywood or hardboard, Z-shaped flashings are often installed. A Z-flashing fits behind the upper sheet and angles over the lower sheet, keeping water from penetrating the joint.

Step flashing is required along the joint between a roof and a dormer or wall that rises above the roof (see drawing above). It's installed during roof installation.

Base line and grading

No matter what type of siding you've chosen, you'll have to align its lowest edge along the base of each wall by snapping a level chalk line no less than 8 inches above grade. (When new siding is going over old, the line is usually set 1 inch below the lower edge of the existing siding.)

If necessary, excavate the landscape where it interferes with this 8-inch clearance, sloping the grade away from the house so that water won't pool by the foundation.

If you have no helper, run the chalk line from a concrete nail pounded into the foundation wall (one at each end if necessary). Be sure the chalk line is level before snapping it.

You may find it necessary to "step" the siding to conform to a hillside or irregular grade (see illustration). If you'll be applying horizontal siding, shingles, or shakes, work out the sizes of any needed steps so that they correspond to the planned exposure for courses of shingles or shakes, panels, or boards.

Base chalk line

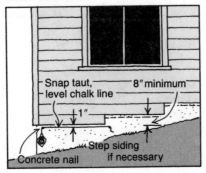

Snap taut, level chalk line

8" minimum

1"

Step siding if necessary

Concrete nail

INSTALLING WOOD-BOARD SIDINGS

For the sake of simplicity in sorting through the many available siding materials, we have grouped all wood-board patterns into a single category. But when it comes to proper installation techniques, wood-board patterns must be viewed individually. They vary greatly.

The chart on pages 110–111 is based on a thorough consideration of each standard pattern. It will tell you whether a particular pattern is applied vertically or horizontally, the type of backing it requires, the proper size of nail to use, and the right nailing technique.

With board siding, controlling water penetration and vapor in a wall is a critical matter; be sure to read through the information on vapor barriers on page 103.

Before you begin applying the siding, figure out how you want to treat the corners. Typical treatments for both inside and outside corners are shown in the drawings on the next page.

When planning the layout of horizontal siding, you may be able to adjust the base line up or down slightly in order to make the layout of boards come out evenly below windows and at soffits. If you can, do so.

Tools you will need: Carpenter's level, caulking gun, chalk line, circular saw, claw hammer, combination square, framing square, ladder or scaffolding, line level, nailset, painting or finishing tools (page 101), plane, saber saw or keyhole saw, safety goggles, T-bevel, tape measure, tin snips, sawhorses, and a utility knife. A few others may come in handy along the way; but with these, you'll be well along your way.

Nailing requires corrosion-resistant nails. Spiral or ring-shank nails offer better holding power than nails with smooth shanks; they work especially well in re-siding over an existing wall covering.

The nail sizes recommended in the chart on page 111 are for new construction. When siding over an existing wall, use nails that penetrate studs at least 1 inch. Use finishing nails where you wish to countersink and fill over nailheads.

If your boards are so thin or dry that they tend to split, predrill nail holes, especially at board ends. Blunting the tips of nails with a hammer also helps keep them from splitting boards.

Corner treatments for wood-board sidings

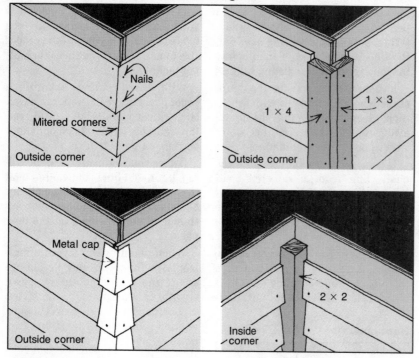

Nails

Mitered corners

Outside corner

1 × 4   1 × 3

Outside corner

Metal cap

Outside corner

2 × 2

Inside corner

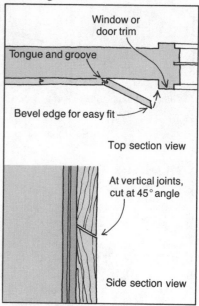

Window or door trim

Tongue and groove

Bevel edge for easy fit

Top section view

At vertical joints, cut at 45° angle

Side section view

Transferring an angle

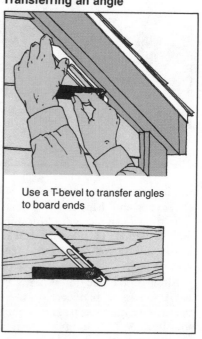

Use a T-bevel to transfer angles to board ends

The first board of horizontal siding goes at the bottom. Beveled types require a starter strip beneath the board's lower edge, along the wall's base, to push it out to match the angle of the other boards (see illustration).

Starter strip

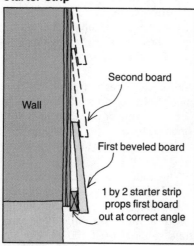

Second board

Wall

First beveled board

1 by 2 starter strip props first board out at correct angle

For vertical wood siding, begin at one corner of the house. Align one edge of the first board with the corner and check its other edge for plumb. If it isn't plumb, trim it with a plane or saw until it is. Be sure the board's lower end is flush with your base chalk line (page 107), then nail it in place.

Successive boards for horizontal siding can be laid out with a "story pole," as described on the next page. Apply the boards from bottom to top. Unless the chart calls for spacing, fit the boards tightly together.

If the wood isn't kiln dried, anticipate some shrinkage as it dries. Consider nailing with just one nail per bearing (stud, furring strip, or blocking) and, if specified, adding the second nail later, when the boards have thoroughly dried.

Installation tips. To fit a shiplap or tongue-and-groove board into a corner against a door or window frame, first rip it to the proper width. Then bevel the edge as shown above right; this will make it easier to push the board into place. Nail.

Where vertical boards join end to end, cut the board ends at a 45° angle as shown to ensure proper water runoff. Waterproof the ends of all boards by brushing on a sealant.

To match the angle of a roofline, measure the angle with a T-bevel. Transfer the angle to board ends, then cut (see drawing at right).

At dormer windows, flash along the roofline as shown on page 107.

Cornices are often left "open" with wood-board siding—the boards extend to the rafters as shown. Trim along the top edge, at the rafters, with quarter-round molding or a narrow trim board (called a "frieze board").

If you prefer a "closed" cornice, you can buy a wood board for the fascia called a "plowed fascia board." This board, which is nailed over the rafter ends, has a routed groove near one edge for holding soffit boards or panels. Build a closed cornice as shown in the drawing below.

Open and closed cornices

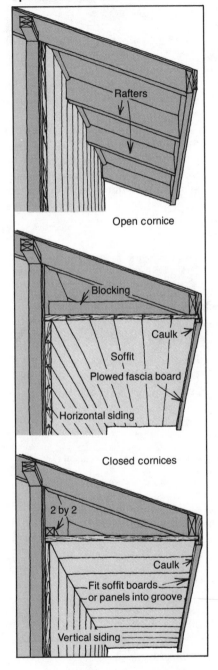

Open cornice

Closed cornices

APPLYING SHINGLES AND SHAKES

The small, manageable size of shingles and shakes makes them one of the easiest of all siding materials to put on.

The backing required for shingles or shakes is a fairly flat, sturdy nailing base. You can also apply them over existing siding; they ride over slightly bumpy wall surfaces better than most siding materials.

Tools you'll need are, for the most part, the same as those listed under wood-board sidings on page 107.

Exposure and coursing. Before applying shingles or shakes, you must determine the correct exposure and whether you will be single coursing or double coursing. The exposure is the amount of each shingle or shake exposed to the weather. Coursing is the method of applying shingles or shakes (see drawings at right above). Though single coursing is by far the most common, double coursing allows longer exposures and the use of low-grade shingles as an undercourse.

First determine the maximum exposure according to this chart:

Maximum Exposures

	Lengths	Single Courses	Double Courses
Shingles	16″	7½″	12″
	18″	8½″	14″
	24″	11½″	16″
Shakes	18″	8½″	14″
	24″	11½″	18″

Now, assuming that you've prepared the wall, as described on page 104, measure the distance from the base chalk line to the soffit at both ends of the tallest wall. Compensating for any steps in the base chalk line, and split any difference to figure the average distance from the soffit to the base line.

Divide that distance by the maximum exposure for your shingles or shakes. Does your computation yield a whole number of courses? Probably not. If not, decrease the exposure enough to make the courses come out evenly. For the neatest appearance, also adjust the number of courses a little to achieve a full exposure below windows.

Single and double coursing

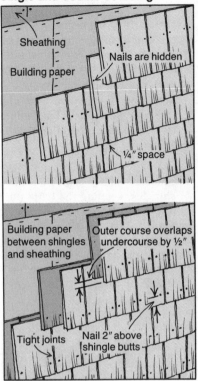

Using a story pole

Make a story pole from a 1 by 2 that is as long as your tallest wall's height (unless that wall is more than one story). Starting at one end, mark the story pole at intervals equaling the established exposure.

Holding or tacking the story pole flush with the base line, transfer the marks to each corner and to the trim at each window and door casing.

Note that, if you plan to double course, shingle or shake butts on the outer course should overlap the undercourse by ½ inch.

First course. Nail on the first course of shingles, keeping the butts flush with the base chalk line (or ½ inch above it if you're double coursing). If you wish, you can use low-grade shingles for the first course whether you're single or double coursing—they will be covered up by the next

course in any case. Allow room for expansion between shingles or shakes: ¼ inch between shingles, ½ inch between shakes.

Directly over the first course, apply a second course. If double coursing, overlap the original course by ½ inch. Offset all joints between

shingles and shakes of different courses by at least 1½ inches so that water will run off properly.

Successive courses. Lay successive courses from the first course to the soffit. As you finish each course, snap a chalk line over it or nail on

Shingling the first course

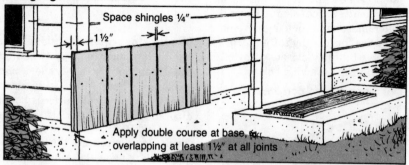

Space shingles ¼"
1½"
Apply double course at base, overlapping at least 1½" at all joints

WOOD-BOARD SIDING: TYPES AND INSTALLATION

	Siding Type	Characteristics	Direction of Application
Board-on-Board — 1" — ½" space — Board-and-Batten	**Board-on-Board** (unmilled)	Available in all standard board sizes, surfaced or rough.	Vertical
Bevel — 1" — Clapboard	**Board-and-Batten** (unmilled)	Available in all standard board sizes, surfaced or rough.	Vertical
Allow ⅛" — Dolly Varden — Shiplap	**Clapboard** (unmilled)	Available in all standard board sizes, surfaced or rough.	Horizontal
Allow ⅛" — Channel Rustic	**Bevel, Dolly Varden**	Plain bevel available with one face smooth, the other sawn. Either may be exposed. Dolly Varden boards are thicker, with rabbeted lower edge.	Horizontal
Allow ⅛" — Drop	**Shiplap, Channel Rustic, Drop**	These overlapping styles are available in a wide variety of patterns and styles.	Horizontal or vertical
	Log Cabin	Standard thickness is 1½" at the thickest point.	Horizontal
Log Cabin — Tongue-and-Groove	**Tongue-and-Groove**	Available with either smooth or rough surfaces; square edges, V-edges, or eased edges.	Vertical, horizontal, or diagonal. Can mix widths.

a straight 1 by 4 board as a guide for laying the next course of shingles or shakes evenly.

If you double course, use a ship-lap board to align the butts of the undercourse and outer course, as shown in the illustration on the next page.

Nails are concealed in single coursing, 1 inch above the line where the butts of the next higher course will go. For shingles, drive a nail ¾ inch in from each side—1 inch for shakes—and then additional nails every 4 inches between.

Shingle nails should be corrosion resistant, 1¼ inches or longer for shingles, 2 inches or longer for

Laying successive courses

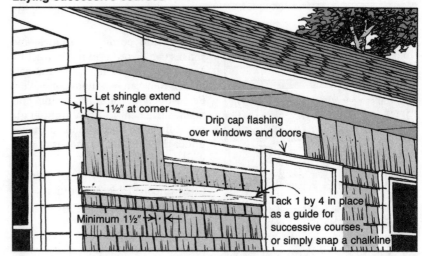

Let shingle extend 1½" at corner

Drip cap flashing over windows and doors

Minimum 1½"

Tack 1 by 4 in place as a guide for successive courses, or simply snap a chalkline

shakes. They should be long enough to penetrate sheathing or solid backing by at least 1 inch. Ring or annular-shank nails hold the best.

Backing Required	Nail Size (For New Construction)	Nailing Tips
For new construction, sheathing may be required. Otherwise, install blocks between studs on 24″ centers. Can be applied directly over all but synthetic sidings and masonry. See pages 104–105.	8d for underboards, 10d for overboards	Face-nail underboards once every 24″ vertically; face-nail overboards twice, 3″ to 4″ apart, at center. Minimum overlap: 1″.
For new construction, sheathing may be required. Otherwise, install blocks between studs on 24″ centers. Can be applied directly over all but synthetic sidings and masonry. See pages 104–105.	8d for underboards, 8d or 10d for battens	Space underboards ½″ apart. Face-nail each board once every 24″ vertically. Minimum overlap: 1″.
Flat surface—sheathing, exposed studs, or furring strips. See pages 104–105.	10d	Face-nail 1″ from overlapping edge (just above preceding course) once per bearing. Minimum overlap: 1″. First board requires wood spacer to prop out at correct angle.
Flat surface—sheathing, exposed studs, or furring strips. See pages 104–105.	8d for ¾″, 6d for thinner	Face-nail once per bearing. With Dolly Varden, face-nail 1″ from lower edge. Allow expansion clearance of ⅛″. Minimum overlap: 1″. First board requires wood spacer to prop out at correct angle.
For horizontal, a flat surface—sheathing, exposed studs, or furring strips. For vertical, install blocks between studs on 24″ centers in new construction, or apply directly over all but synthetic sidings and masonry. See pages 104–105.	8d for 1″, 6d for thinner	Face-nail once per bearing for 6″ widths, twice (about 1″ from overlapping edges) for wider styles.
Flat surface—sheathing, exposed studs, or furring strips. See pages 104–105.	10d	Face-nail once per bearing for 6″ widths, twice for wider styles.
For horizontal, a flat surface—sheathing, exposed studs, or furring strips. For vertical or diagonal, install blocks between studs on 24″ centers in new construction, or apply directly over all but synthetic sidings and masonry. See pages 104–105.	8d (finishing nails for blind-nailing, otherwise 8d siding nails)	Blind-nail 4″ to 6″ widths through tongue with finishing nails, once per bearing. Face-nail wider boards with two siding nails per bearing.

Aligning double courses

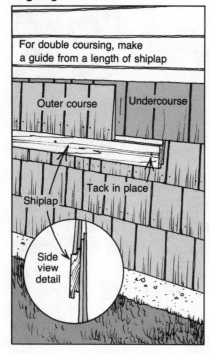

For double coursing, make a guide from a length of shiplap

Outer course

Undercourse

Shiplap

Tack in place

Side view detail

For double courses, place the nails 2 inches above the butt line, ¾ inch in from each edge, and at 4-inch intervals in between.

Corners. Typical methods for finishing shingles at corners are shown below. For outside corners, the shingles or shakes can be brought flush to a vertical 1 by 3 or 1 by 4. Or you can miter the corners, but be prepared to spend a lot of time at it.

More typically, you can "weave" the corners by alternately overlapping them.

At inside corners, shingles and shakes can be brought flush to a 1 by 1 or 2 by 2 board nailed in the corner. Or they can be mitered or woven. For woven or mitered inside corners, flash behind with right-angle metal flashing that extends 3 inches under the shingles or shakes of each wall.

Caulk the seams well at corners. Clear silicone looks better than white latex caulk, but will not take a stain or paint as well as latex.

Obstacles. Shingles and shakes are easy to cut and fit around obstacles such as doors, windows, pipes, meters, and so forth. For curving cuts, use a coping saw or saber saw. Caulk well around obstacles.

Cornices can be handled in either "open" or "closed" fashion. You can trim shingles or shakes to fit around rafters for open cornice treatment. Closed cornices are best handled with tongue-and-groove or shiplap wood-board siding, as described on page 109.

INSTALLING PLYWOOD

As shown on pages 88–89, there are two main categories of plywood siding on the market: plywood sheets and lap-type panels. Plywood sheet siding, one of the easiest sidings to install, goes up fast, covering large areas rapidly. Lap panels are installed in almost the same way as the horizontal wood-board patterns they mimic, with a few variations.

The following directions will tell you how to install plywood sheet siding. For lap panel installation techniques, refer to page 107 where wood boards are discussed. To supplement that information, a few specific tips for applying plywood lap panels accompany the directions here.

The backing required. Plywood sheets may be applied directly over studs, flat existing siding, or furring strips. Sheathing is not required behind plywood sheet siding, but it does add to a wall's rigidity when the siding sheets are ⅜ inch thick or less. Building paper is needed only behind those vertical joints between panels that neither interlock nor are covered by battens. It adds an extra barrier against air and moisture.

Sheets are installed either vertically or horizontally. Because vertical installation minimizes the number of horizontal joints, it is the most common method.

For the proper layout of studs, blocking, or furring behind plywood sheets, see page 105.

Typical corner treatments for shingles

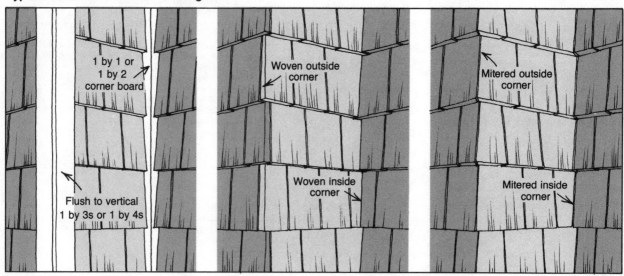

1 by 1 or 1 by 2 corner board

Flush to vertical 1 by 3s or 1 by 4s

Woven outside corner

Woven inside corner

Mitered outside corner

Mitered inside corner

Plywood lap panels usually require a backing of sheathing to give a wall strength (unless you apply them over an existing wall). Building paper is also needed. Check the manufacturer's recommendations and local codes to see what's required.

Tools you'll need are the same as for wood-board siding application, listed on page 107.

The first step. Before you begin putting on plywood sheets, you'll need to figure their proper lengths. They should reach from the base chalk line (page 107) to the soffit. If you plan to create a closed soffit, leave enough space for fitting the soffit boards or panels in place above the siding (refer to the information on cornice treatments on page 115).

Plywood horizontal joints

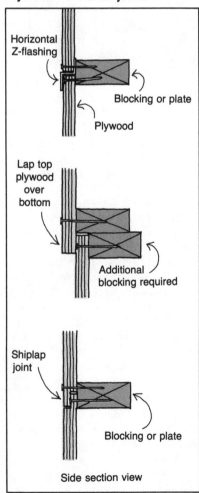

Horizontal Z-flashing

Blocking or plate

Plywood

Lap top plywood over bottom

Additional blocking required

Shiplap joint

Blocking or plate

Side section view

If the distance from base chalk line to soffit is longer than the plywood sheets, you'll need to join panels end to end, using one of the methods for horizontal joints shown in the drawings, below left. If you use Z-flashing, be sure it will keep water from puddling. Back the panel edges at horizontal joints with blocking or other firm nailing support.

Next, determine where the studs are, if they are hidden behind a wall. Look for nails in vertical rows, or probe for the studs by drilling small holes. The studs should occur every 16 or 24 inches, from center to center, and on both sides of windows and doors. Mark their locations along the foundation wall and above the area the siding will cover (use a plumb bob to align top and bottom marks).

The first sheet is mounted at an outside corner, its bottom edge flush with the base chalk line. Check with a level to make sure vertical edges are plumb. If the corner itself isn't plumb, you'll need to trim the plywood edge to align with it, or else taper corner moldings later on to adjust the alignment. The best method will depend on where the wall studs lie—the vertical plywood edge opposite the corner must be aligned with a stud—and the corner treatment you prefer. See the information on corner treatments that begins on page 114.

Trimming the edge of the plywood sheet to align with the corner is easy. Hold or tack the sheet in place, flush with the base chalk line. The inside vertical edge must align with a stud, furring strip, or other firm backing. Mark—or have a helper mark—the edge as shown, along the outermost points of existing siding or framing, at top and bottom.

Take the panel down and snap a chalk line between the two marks. Cut along the line, using a circular saw. Nail the trimmed panel in place as discussed on page 114.

Trimming edge of first sheet

Flush at eave

Align edge with stud

Mark for cutting

Panel's edge

Back lap

Front lap

Overlap old siding by 1"

The successive sheet butts against the first sheet, often with an overlapping shiplap vertical edge (see detail in drawing at right). Leave a $1/16$-inch expansion gap at all joints (in humid climates, leave $1/8$ inch). Sheets must join over studs, blocking, or some other type of sturdy backing. Don't nail through laps, but as shown in the detail.

If your plywood doesn't have a shiplap edge, caulk along the vertical edges and butt them loosely together, leaving about $1/16$ inch. Then cover the joints with battens. (You can butt the sheets without battens if you put building paper behind each joint.)

As you lay sheets, you may encounter obstacles such as gas pipes

Plywood vertical joints

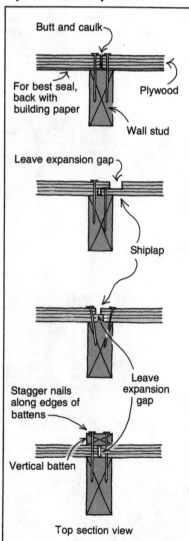

Butt and caulk

For best seal, back with building paper

Plywood

Wall stud

Leave expansion gap

Shiplap

Leave expansion gap

Stagger nails along edges of battens

Vertical batten

Top section view

Putting up successive sheets

Wall stud

Old siding

Sheathing

Back lap

Front lap

Cut out slot with saber saw

Metal drip edge

Leave $3/16''$ gap around windows for caulking

Cut piece to fit around protrusion, then glue in place

or hose bibs. Cut out a section with a saber saw, so that the sheet can be put in place (see detail in drawing above). Mount and nail the sheet; then fashion the cutout so it will fit around the protrusion and fasten it in place with glue and—if possible—nails.

Nail plywood sheets with corrosion-resistant nails. For re-siding over wood boards or sheathing, use hot-dipped galvanized ring-shanked nails. Do not use finishing nails for nailing up plywood siding.

Nails should be long enough to penetrate studs or other backing by $1\frac{1}{2}$ inches. Nail every 6 inches around the perimeter of each plywood sheet and every 12 inches where the sheet crosses studs or other backing members. Avoid making a "dimple" in the plywood surface with the last hammer blow when pounding in nails.

Corners for plywood walls require special attention to ensure weather-tightness. Where there is prolonged contact with water at a corner, delamination can take place.

Outside corners can be rabbeted together and caulked or covered by a 1 by 3 and 1 by 4 trim board, as shown. Inside corners are generally just caulked and butted together. You can also apply a vertical corner trim board, such as a 2 by 2, to inside corners (see drawings on facing page).

If the plywood is milled with grooves, the grooves under vertical corner boards might let dirt and water penetrate, causing rot. To prevent this, nail the vertical corner boards directly to the corner of the wall or corner studs, caulk liberally along the inside edges, and butt the plywood sheets against them.

Windows and doors should have metal drip caps, as discussed on page 106. Be sure that drip caps are sloped slightly downward for proper runoff.

When cutting out the large areas necessary for windows and doors,

Outside corners

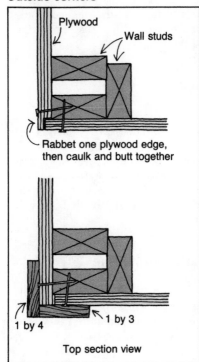

Plywood

Wall studs

Rabbet one plywood edge, then caulk and butt together

1 by 4   1 by 3

Top section view

Inside corners

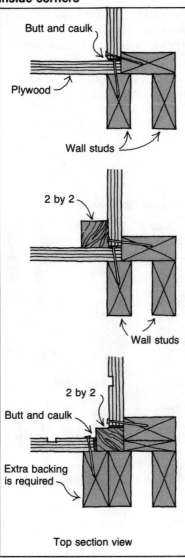

Butt and caulk

Plywood

Wall studs

2 by 2

Wall studs

2 by 2

Butt and caulk

Extra backing is required

Top section view

Plywood cornices

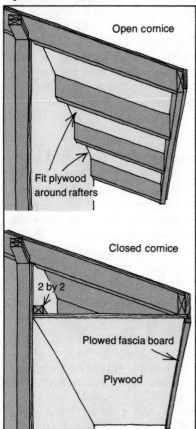

Open cornice

Fit plywood around rafters

Closed cornice

2 by 2

Plowed fascia board

Plywood

Plywood lap installation

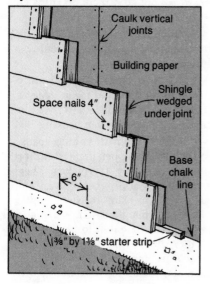

Caulk vertical joints

Building paper

Shingle wedged under joint

Space nails 4"

Base chalk line

6"

⅜" by 1⅜" starter strip

---

abide by the wisdom of the carpenter's maxim, "Measure twice, cut once." Use a straightedge or chalk line to mark cuts, a circular saw for cutting straight, and a saber saw for cutting corners and curves.

Cornices on plywood-sided houses are usually closed. An open cornice requires very exacting cuts to fit snugly around the rafters. Consider running a plowed fascia board along rafter tails and a piece of plywood siding along the soffit, as shown above, far right.

To do this, mark a point on the wall level with the top of the groove in the fascia board, using a level. Find this point at both ends of the wall, then snap a chalk line between your two marks. Nail a 2 by 2 above the chalk line. Measure the distance from the interior of the plowed fascia groove to the wall. Cut the soffit board to fit, and cut sufficient holes in it for attic ventilation (see page 63). Install the soffit board in the plowed fascia and nail it every foot to the bottom of the 2 by 2. Hide the nails with quarter round or other molding, fastening it down with finishing nails.

Installing plywood lap siding. Refer to the section on wood-board siding installation on page 107 for general techniques. Here are a few specific tips for plywood lap siding (see drawings):

Install a ⅜-inch by 1⅜-inch starter board along the base chalk line (page 107) to keep the lowest lap panel at the same angle as successive panels.

Where vertical joints occur in plywood lap siding, insert a shingle wedge under the joint to support the wood.

Treat the corners with water sealer in advance, and caulk all joints during application. For most types of plywood siding panels, joints may occur away from studs if applied over nailable sheathing (check manufacturer's recommendations).

Nail at joints with backing and every 16 or 24 inches along the lap siding, depending on the spacing of backing. Nails should penetrate studs or other solid backing by 1½ inches. Nail the vertical edges of panels every 4 inches. Nail along the bottom of the starter panel into the starter strip every 6 inches. At intermediate 24-inch nailing points, use one nail per panel or nail every 8 inches vertically for wide panels.

INSTALLING HARDBOARD

The application of hardboard parallels that of plywood in most respects (see page 112). However, there are a few differences.

Because hardboard is not as strong as plywood, more restrictions govern its use. Check your local building codes for the regulations in your area.

Hardboard sheets in new construction may generally be applied directly to studs spaced no more than 16 inches apart, center to center. Some hardboard patterns can be put directly over studs 24 inches apart.

Building paper is usually required under hardboard sheets if the hardboard is applied directly to studs or wood-board sheathing.

As with plywood sheets, all joints between hardboard panels should be located over studs.

Hardboard sheets are generally nailed around the perimeter every 4 inches and intermediately every 8 inches at studs or other backing.

Use box nails with ¼-inch heads, long enough to penetrate studs or other backing by 1½ inches. They must be hot-dipped galvanized or other corrosion-resistant nails. Hardboard manufacturers sell nails with heads colored to match the factory finish on the hardboard. Matching colored nails (as well as matching caulking compound) can save you time in finishing later.

When nailing, be careful not to dimple the hardboard by pounding the nails in too far, especially when nailing over foam insulation board.

Hardboard lap siding must go over sheathing. Otherwise, the rela-

tively narrow hardboard panels can take on a wavy appearance. Also use building paper.

Always join panels over studs. Stagger joints so they don't line up.

Nail lapped hardboard panels with two nails, one ⅜ inch from the butt of the section above and the other ¾ inch from the bottom edge. Unlike the nails used with wooden bevel boards, the nails for hardboard penetrate two layers of material.

Hardboard lap installation

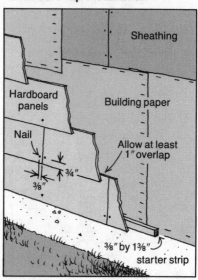

Sheathing

Building paper

Hardboard panels

Nail

Allow at least 1″ overlap

¾″

⅜″

⅜″ by 1⅜″ starter strip

INSTALLING VINYL SIDING

Panels of vinyl siding, applied either horizontally or vertically to house walls, are part of a siding system that also includes assorted trim (see drawings on next page). The trim is installed first—around the house base line and window and door frames, and against soffit or gable edges—and panels are then fitted into it for a uniform, finished appearance.

The backing you'll need

The backing for vinyl siding must be smooth enough for the panels to lie flat. For such a surface you'll need

either smooth sheathing over wall studs or furring strips nailed to the butts of lap siding. See "Preparing the Wall," page 104.

Tools

To work with vinyl, you'll need a radial or circular saw, safety glasses (to wear when using the saw), work table, hammer, rule, level, snap-lock punch, tinsnips or aviation shears, scoring tool, utility knife, carpenter's square, and plumb bob with line.

Nails and nailing

With vinyl siding use only aluminum or corrosion-resistant box nails long enough to penetrate 1 inch into the studs or nailing base.

The two most important things to remember when nailing vinyl panels are: 1) unless otherwise specified, always nail in the center of nailing slots; and 2) never drive nails in too tight.

How to cut vinyl

When cutting vinyl panels, it's easiest to use a radial arm saw or circular saw equipped with a plywood or fine-tooth blade (12 to 16 teeth per inch). For some cuts, tinsnips or aviation shears are used; for others, a utility knife. Corner posts are best cut with a hacksaw. (See drawings below right.)

Installing corner posts

Install corner posts at inside and outside corners, allowing ¼ inch for expansion at the upper trim line and running them down to the base chalkline (page 107).

Because vinyl expands and contracts excessively in heat and cold, the full ¼ inch is necessary at all joints and at ends of panels. (For aluminum, which expands less, you would leave ⅛ inch; for steel, you would leave a 1/16 inch.)

Vinyl, aluminum, and steel siding components

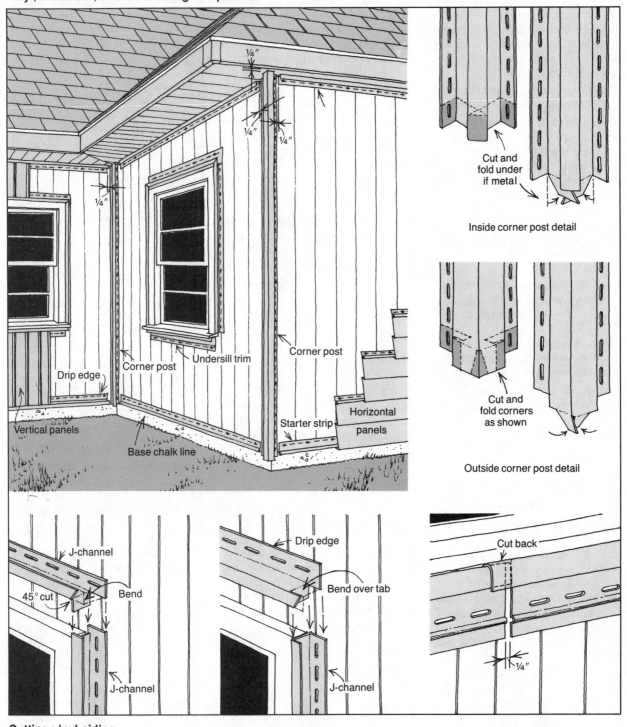

¼"
¼"
¼"
¼"

Corner post
Undersill trim
Corner post
Drip edge
Horizontal panels
Starter strip
Vertical panels
Base chalk line

Cut and fold under if metal

Inside corner post detail

Cut and fold corners as shown

Outside corner post detail

J-channel
45° cut
Bend
J-channel

Drip edge
Bend over tab
J-channel

Cut back
¼"

Cutting vinyl siding

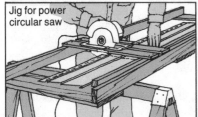

Jig for power circular saw

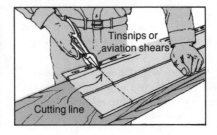

Tinsnips or aviation shears
Cutting line

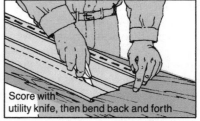

Score with utility knife, then bend back and forth

Position each post by driving two nails through the top of the upper-most slots; the post should hang from the nails. Then use a level to check for vertical alignment, and make necessary adjustments.

Fasten the posts to the wall with nails positioned every 12 inches.

If you have to stack one piece of corner post above another, trim ¼ inch from the nailing flange at the bottom end of the top post. Then mount the top post so it overlaps the lower one by 1 inch.

Installing the starter strip or drip cap

Install the starter strip (for horizontal siding) or drip cap (for vertical siding) along the base chalk line.

Standard application. Align the upper edge of the base trim with the chalk line and nail every 6 inches.

When you come to an outside corner, allow ¼ inch for expansion between the starter strip or drip cap and corner post.

Installing door and window trim

Your next step is to mount strips of appropriate trim—usually J-channel, drip edge, or undersill—around window and door openings. To reduce moisture and air flow, however, first run a bead of caulk around the openings to provide a seal. (If you intend to put vinyl covers on window sills and casings, do that before installing accessory trim.)

When you install door and window trim, first do the top, then the sides, and finally the undersill trim.

Installing top trim. Along tops of doors and windows, install J-channel trim for horizontal siding and drip edges for vertical panels. Use lengths of trim that measure two channel widths longer than the top of the opening; either miter the ends or cut tabs at each end. Nail 12 inches on center.

Installing side trim. Along the sides, use mitered trim if the top trim is mitered. When fitting side pieces, position the top nail at the top of the nailing slot, but drive remaining nails every 12 inches in the slot centers.

To avoid problems fitting horizontal panels into narrow places—between adjacent windows, for example—fasten down the trim along only one side of such areas. Trim on the other side can be fastened down as panels are put in place.

Installing trim underneath windows. Under windows, install trim for horizontal siding and J-channel trim for vertical. Add furring strips where necessary to maintain the slope of the siding.

Installing trim under eaves and rakes

If you're installing horizontal siding, use undersill trim at the soffit and at the gable rake (remove existing rake molding first). For vertical panels, install J-channel trim. Nail 12 inches on center. Add furring strips if necessary.

Installing horizontal panels

Allowing ¼ inch for expansion at each end where the panel fits into the J-channel, fit and securely lock the starter panel into the starter strip. (If you live in a cold climate, allow ⅜ inch expansion space.)

If you are installing insulative backerboard with the siding, drop it behind the panel, beveled edge down and toward the wall.

Fitting end joints. Where panels meet, overlap the ends one inch. Run overlaps away from the most common vista on each wall—such as the front walk or porch—so that joints will be less obtrusive.

When overlapping, cut 1½ inches of the nailing flange away from the end of one panel to allow for expansion (see drawing above).

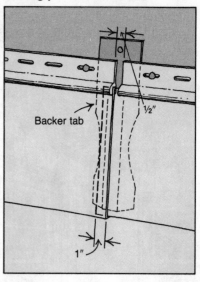

Backer tab ½" 1"

Slip backer tabs, flat side out, behind the joints of 8-inch noninsulated panels to ensure that the joints will stay rigid.

Try to offset joints at least 24 inches from one course to the next so that vertical seams don't lie one above another unless separated by several courses.

Fitting panels around windows and doors. You'll need a sharp knife, tinsnips, or a power saw (for insulated siding) to trim panels to fit around windows and doors, allowing ¼ inch for expansion. Remember, though, to use either the saw or knife—never the tinsnips—to cut the locking detail at the lower edge of the panel.

Measure window and door openings and cut panels accordingly, adding ¼ inch (⅜ inch in cold climates) for expansion. Before fitting cut panels to undersill trim, use a snap-lock punch to crimp nubs or "ears" along the trimmed edge. The nubs, spaced 6 to 8 inches apart, should face outward.

To fit horizontal panels into narrow places, each panel is slipped into the trim along one side; as successive panels are slipped into place, nail down the trim along the other side.

Installing top panels at eaves or gables. Measure from the bottom of the top lock to the eave and subtract ¼ inch (⅜ inch in cold climates) for expansion. To determine the width of the final horizontal panel, measure in several places along the eaves.

Cut the panel as needed, and use the snap-lock punch to crimp nubs along its upper edge every 6 to 8 inches. Then tuck the panel into the trim.

At the gables, cut panels at an angle to fit into the J-channels along the gable rake.

Installing vertical panels

Once the corner posts and trim are in, locate the center of each wall and, using a level and a straightedge, draw a line down the center. Allowing ⅜ inch for expansion at the top of the first panel, center it over the line. Fasten it with nails positioned at the top of the nailing slots every 8 inches.

Working from the starter strip, install successive panels. They should be long enough to fit between the trim strips, minus ⅜ inch for expansion (panels should rest on the drip cap at the base).

Insert each panel first into the J-channel along the top of the wall, then let it rest on the drip edge and lock it into the previous panel.

Nails for successive panels should be positioned in the slot centers, every 8 to 16 inches depending on the manufacturer's recommendations.

When fitting vertical panels around windows and doors, follow the instructions for "Installing horizontal panels," preceding.

Corner finishes. Before you insert the last panels into the corner posts, install J or U-channels or undersill trim in the corner post slots (manufacturers' instructions will vary). You may want to raise the J-channels with ⁵⁄₁₆-inch shims (for ½-inch J-channels) to keep panels on the same plane.

Then, as illustrated above, insert uncut panel edges (and cut edges

Corner treatments

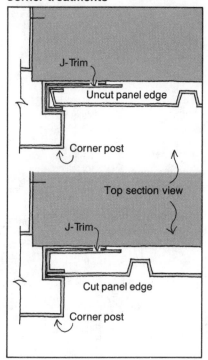

near V-grooves) into the J-channel. Cut edges of flat sections should be inserted between the J-channel and the post's outer flange.

Special situations

To complete the vinyl siding system, manufacturers make fascia trim, window trim, and other specialty parts for a finished look. The installation of these—as well as soffit panel application and the handling of transitions between horizontal and vertical panels—varies from one manufacturer to another. For information on these special products and installation procedures, read manufacturers' recommendations.

INSTALLING ALUMINUM AND STEEL SIDINGS

Aluminum and steel sidings are installed just as vinyl siding is, with these exceptions:

1. A brake, available from tool rental companies, is used to bend aluminum so it conforms to corners and window and fascia trim. The brake is usually also used to make finishing trim, from aluminum coilstock.

2. Gutter seal adhesive, available from your siding supplier, holds aluminum panels under window sills and at soffits, especially where there is no other means of support. The gutter seal serves the same purpose as the ears or nubs made on vinyl with a snap-lock punch.

3. Metal siding fittings are often folded, while vinyl is not. The aluminum J-fittings around windows, for example, are often bent for a tight fit.

4. Backing tabs are used at joints in horizontal aluminum siding to make the aluminum more stable over long spans.

5. Nails for aluminum siding should be aluminum. Nails for steel siding should be galvanized steel, to avoid the corrosion that might result from reactions between different metals.

While all nails in vinyl siding are concealed, some aluminum trim pieces are held with nails that puncture the material.

6. Because metal siding is thinner than vinyl, it may be necessary to use thicker furring strips under windows and soffits to make a uniform plane.

7. The aluminum starter strip for vertical panels may be the same as is used for horizontal panels.

8. While aluminum siding can be cut with tinsnips or a circular saw fitted with a plywood or fine-tooth blade, steel siding must be cut with heavy-duty tinsnips or a special guillotine cutter. The guillotine cutter is recommended because it crimps the galvanized coating over the end of the steel and minimizes rusting. (You can also paint cut corners of steel siding to reduce the risk of rust.)

Take special care when cutting aluminum or steel to wear safety glasses, as well as heavy clothing and gloves—cutting metal with a saw generates metal slivers that can cause injuries.

Index

Boldface numbers refer to color photographs

Protecting Against the Weather

- Install your own roofing and siding, with the help of over 150 how-to illustrations

- Learn how to buy roofing and siding materials, with comparative charts plus guidelines for reliable estimating

- Discover how to use different materials to best advantage—shown in more than 70 color photos

- Select from a complete listing of gutters and downspouts now available, and how to install them

- Diagnose problem situations and take follow-up steps—finding leaks, replacing shingles, making roof repairs, patching siding

Asphalt shingles and weathered spruce siding combine for handsome exterior of traditional house (above). Classic board-and-batten siding (below) adds strong vertical lines to contemporary house.

MORE HOME IMPROVEMENT BOOKS FROM *Sunset*

Barbecue Building Book
Basic Carpentry Illustrated
Basic Home Wiring Illustrated
Basic Masonry Illustrated
Basic Plumbing Illustrated
Decks
Fences & Gates
Fireplaces
Garden & Patio Building Book
Garden Pools, Fountains & Waterfalls
Home Remodeling Illustrated

Home Repair Handbook
Insulation & Weatherstripping
Outdoor Furniture
Patio Roofs & Gazebos
Patios & Decks
Solar Heating & Cooling
Solar Remodeling
Spas, Hot Tubs & Home Saunas
Swimming Pools
Walks, Walls & Patio Floors
Windows & Skylights
Wood Stoves

LANE PUBLISHING CO.,
MENLO PARK,
CALIFORNIA 94025

Publishers of Sunset Books for Building, Gardening, Cooking, Hobbies, and Travel.

$7.95 U.S. $10.95 CANADA
ISBN 0-376-01491-1

0 70661 01491 3

Sunset Tile

REMODELING HANDBOOK
Ceramic • Vinyl • Wood • Step-by-Step Techniques